From Markov Jump Processes to Spatial Queues

From Markov Jump Processes to Spatial Queues

by

Lothar Breuer

Department of Computer Science,
University of Trier, Germany

Springer Science+Business Media, LLC

A C.I.P. Catalogue record for this book is available from the Library of Congress.

DOI 10.1007/978-94-010-0239-4

Printed on acid-free paper

Originally published by Kluwer Academic Publishers in 2003
Softcover reprint of the hardcover 1st edition 2003
MyCopy version of the original edition 2003

Springer.com/mycopy

Contents

Part II Classical Queues

Part III Spatial Queues

Preface

The development of applied mathematics is to a large part driven by the constant flow of technological innovation. The fruitful interrelation between queueing theory and its application field of communication networks is a perfect example for this. One of the major recent challenges for the stochastic modelling of communication systems is the emergence of wireless networks which are used by more and more subscribers today. The main new feature of those, which has not been covered by classical queueing theory, clearly is the importance of the user location within the area that is served by the base stations of the network.

In the framework of queueing theory, this opens up the natural extension of classical queueing models towards queues with a structured space in which users are served. For wireless communication networks, the classical concepts of multi–class users or queueing networks seem awkward and do not provide enough results to reflect all features that are important for their performance analysis.

The present book is intended to introduce the above mentioned extension under the name of spatial queues. The main point of view and the general approach will be that of Markov jump processes. Starting from a closer look into their theory, new results are presented for the theory of stochastic processes as well as for classical queueing theory before the new concepts of spatial Markovian arrival processes and spatial queues are introduced.

Thus the present work not only opens up the fresh field of spatial queues, but also provides substantial new results for time–inhomogeneous Markov jump processes and queues, in the theory of Markov–additive jump processes, and for statistical model fitting of arrival processes.

The book is intended to reach queueing theorists, researchers in the field of communication systems, as well as engineers with some background in probability theory. It further should be suitable as a textbook for advanced queueing theory on the graduate or post–graduate level.

The main text is divided into three parts. The first part provides a new presentation of the theory of Markov jump processes, deriving a number of new results, especially for time–inhomogeneous processes, which have been neglected too much in the current textbooks on stochastic processes. Further, the class of Markov–additive jump processes is analyzed in detail for the first time in this book. This extends and unifies all Markovian arrival processes that have been proposed up to now (including arrivals for fluid queues) and provides a foundation for the later introduction of spatial Markovian arrival processes.

The second part contains new results for classical queues with BMAP input. These include the first explicit formulae for the distribution of periodic queues and a generalization of results for the inhomogeneous $M_t/G/\infty$ queue towards the inhomogeneous $BMAP/G/\infty$ queue. To this aim, the concept of inhomogeneous BMAPs is introduced first. Further the class of fluid Markovian arrival processes is introduced, as well as a statistical estimation for the parameters of a BMAP.

In the third part, the concepts of spatial Markovian arrival processes (abbreviated: SMAPs) and spatial queues are introduced. After that, periodic spatial Markovian queues are analyzed as a model for the cells of a wireless communication network. The class of $SMAP/G/\infty$ queues is examined, and it is demonstrated how to use them as a planning tool for the design of mobile communication networks. Finally, for the class of homogeneous SMAPs with finitely many phases a statistical procedure for model fitting is presented.

For readers only interested in the new results for the queueing systems treated in this work, the first part may be skipped at first reading. Researchers who are only interested in the spatial concepts and already familiar with the phase method of BMAPs and PH distributions can skip the second part of the book. The extensive presentation in this part is intended to show the use of time–inhomogeneous Markov jump processes for classical queueing theory, but also to make the introduction to a theory of spatial queues more comprehensible.

Parts II and III are organized and written in almost exactly the same schemes. This shall help to show the strong analogy between the treatments of spatial and classical queues. In fact, these are merely two different lines of the same development starting from different specifications of the class of Markov–additive jump processes introduced in the first part of the book.

This work originates from a PhD thesis written at the University of Trier, Germany, during the years 1998 through 2000 (see Breuer [31]). This has been reorganized fundamentally in order to render it accessible for the greatest possible audience. Chapters 2, 4, 5, 6, and 8 have been published before in Breuer [33, 32, 35, 28, 29], respectively. The author hopes that the goal of disseminating the new concepts contained in this book as broadly as possible will not be futile.

Trier, September 2002

Acknowledgments

The fundamental idea of how to construct Spatial Markovian Arrival Processes goes back to Prof. Dieter Baum, who thus inspired the topic of the present work. Through many helpful discussions, he continued to motivate and support this work during its process of development. For this, I would like to express my gratitude and appreciation. Further, I would like to thank Prof. Igor Kovalenko for his helpful comments, which led to substantial improvements in the final version of this work. Oleg Gaier helped me designing the figures in the book. Last but not least, I would like to appreciate the patience of my friend Corinna, who had to suffer many times of absentmindedness and solitary mood, which seem to be characteristic byproducts of mathematical research.

Introduction

The common gains from the interrelation between mathematical theory and practical application can be traced in the field of queueing theory especially well. Starting with Erlang's research at the beginning of the 20th century, queueing theory was born as the endeavour of building appropriate stochastic models for the performance analysis of technical systems. The initial application area of telephone systems was soon enlarged by applications in machine repair, inventory control, assessment of insurance risk and later the design and analysis of computer systems, to name but a few. The close interaction between theory and practice has remained a driving force for the development of queueing theory until today.

In this respect, the momentum coming from applicational demands in the field of telephone networks has not lost its strength. The integrated transmission of voice, data and video sources via the same channel (e.g. in ISDN systems) crucially contributed to the shift from telephone networks of Erlang's times towards telecommunication systems of today. This required the use of more sophisticated models for data streams than the now already classical Poisson process could provide. At the end of the 1970s, Neuts [95] proposed a richer class of processes, which under the name "Batch Markovian Arrival Processes" (shortly BMAPs) was notationally simplified by Lucantoni [80] in 1991. This class of processes, which actually is a subset of the class of Markov–additive processes introduced by Çinlar [41, 42] in 1972 and even earlier by Ezhov and Skorokhod [55, 56] in 1969 or Neveu [99] in 1961, engendered a new direction in queueing theory using so–called matrix–analytical methods.

The extension presented in this work is motivated by a further step in the development of modern telecommunication systems. The technical availability of wireless data transmission opened up the way for mobile communication systems, which today are used by increasingly many subscribers. At the same time, this development exhibited the need for queueing models that can reflect the spatial distribution of the users in the system. These are indispensable in order to model phenomena like spatially dependent service time distributions, user movements or the spatial distribution of users in the system. The analysis of spatial queues yields such essential values as the necessary cell capacities needed to guarantee a low outage probability, or the outage probability of an existing network. Such knowledge greatly improves the competence of planning and operating mobile communication networks.

There have been several authors who introduced queueing models designed to explore the spatial properties of mobile communication systems. Some of the recent notable contributions are the following:

In 1995, Çinlar [44] used the concept of Poisson random measures in order to obtain distributions for the spatial $M/G/\infty$ queue with Poisson arrivals in time and space. Furthermore, mobility of customers is modelled using a method from Massey and Whitt [84], namely integrating the stochastic location process with respect to the Poisson arrival process. The same approach will be used in chapter 9 of this book for a more general model.

Two years later, Baccelli and Zuyev [12] introduced a sophisticated model for mobile communication networks using mainly concepts from the theory of point processes. Analytical results are given only for models which are based on the Poisson process. One of the objectives of the present work is to obtain explicit results for models more general than Poisson models.

The same authors have contributed useful results from stochastic geometry, see Baccelli et al. [11, 10]. A new kind of stochastic point fields has recently been developed by Latouche and Remiche [109, 111, 110]. Further new results on stochastic point fields can be found amply, e.g. Foss and Zujev [60], Chin and Baddeley [38, 39], or Mecke and Stoyan [86], to name but a few. These approaches do of course not provide a temporal dimension.

Boucherie, Dijk and Mandjes [124, 25] developed a model for cellular mobile communication networks which is based on Jackson networks and more general product form networks. Thus the analysis can be performed using known methods for queueing networks. The spatial distribution has a granularity as fine as the number of cells in the network. For further results from this approach see Boucherie and Dijk [23, 24, 21, 22].

Spatial generalizations of the classical $M/G/\infty$ queue as well as several variants of spatial queueing networks can be found in chapters 9 and 10 of a monograph by Serfozo [119] that appeared in 1999. In this, the general approach comes from the theory of point processes.

The crucial ideas for large parts of the concepts presented here came from Baum [17, 35]. In these papers, a concept of spatial BMAPs is developed and the respective infinite server queue is analyzed. It is shown that this spatial generalization retains the properties of BMAPs that yield a tractable analysis of such arrival processes and their respective queues. These ideas are taken up in this work and lead to the construction of the Spatial Markovian Arrival

Process (SMAP). It turns out that the analytical methods developed in Baum [16, 35], which were intended to be applied to queues with BMAP input, can be adapted to the analysis of several practically relevant queues with SMAP input.

The aim of the present work in this context is to develop a unified theory of spatial queues, which is broad enough to serve as a base for queueing models arising from applicational needs in any area involving a spatial dimension of the examined system, but also specific enough to yield concrete results for the performance analysis of mobile communication networks. A particular objective is to develop the most natural generalization of existing concepts (e.g. the BMAP) toward the needs of mobile communication networks. To these belong the spatial distribution of batch arrivals and users in the system as well as time–inhomogeneous (e.g. periodic) arrival intensities and user movements. The concepts used in this work will be embedded into existing mathematical concepts. Thus it is possible to show the similarity of those to concepts used in classical queueing theory.

The book is divided into three parts. In part I, the main theoretical concepts which are used in this work are introduced. This involves the theory of general (i.e. possibly inhomogeneous) Markov jump processes and the concept of Markov–additive jump processes. For readers who are only interested in results for the analyzed queueing systems, this part can be neglected during a first reading. However, all the methods of analysis are finally only applications of results which are proven in this part of the book.

Chapter 1 contains a new treatment of general Markov jump processes (abbreviated: MJPs). Going beyond all current textbook presentations, Markov jump processes are examined in their full generality with general state spaces and time–dependent infinitesimal rates. For general Markov processes inhomogeneity in time does not yield a substantial generalization. But for Markov jump processes an explicit examination of the inhomogeneous case preserves the pure jump structure of the sample paths and thus leads to substantially new results. After a specification of the homogeneous case and an approximation of general Markov jump processes by piecewise homogeneous MJPs, two special cases are examined on their own. The first is that of MJPs with a periodic generator function, while in the second case the structure of the generator function allows a much simpler representation of the transition kernels.

In mobile communication networks the integrated transmission of different data streams is of standard use. Therefore, it seems appropriate to base models of such networks on the BMAP concept in order to preserve the capability of modelling integrated service transmission. In the present work, before application–oriented models are developed in terms of spatial queues, the concept of BMAPs is generalized towards a special class of Markov jump processes in chapter 2. This class will be called Markov–additive jump processes. For them, several properties are derived which allow the same methods of analysis to be used for spatial queues (as set forth in this work) as well as for the very general class of queues on real vector spaces (instead of the classical queues on the space of non–negative integers). This opens the way for the application of queueing theory to many fields which cannot be modelled by classical one–dimensional queues. Besides the extension of applicability of the theory, this chapter intends to exhibit fundamental results on arrival processes on a fairly abstract level in order to further explore the nature of Markov–additive jump processes as an own interest.

Part II contains an application of some of the new concepts obtained in the first part to classical, i.e. non–spatial queues. This shall lead to a better understanding of the methods of analysis in the classical framework, before they are applied to the new field of spatial queues. The reader who is only interested in spatial queues and confident enough to understand the methods of analysis without analogies to non–spatial queues can skip this part of the book.

In chapter 3 the class of BMAPs is extended toward time–dependent arrival rates. Further a new class of fluid Markovian arival processes (FMAPs) is introduced. Both are mere specifications of the class of Markov–additive jump processes as introduced in part I of the book. The presentation shall show the structural similarities between them and encourage the interested reader to perform the same analysis that will be explicated for queues with BMAP input to fluid queues, i.e. queues with FMAP input.

Chapter 4 contains an analysis of the periodic $BMAP/PH/c$ queue. This queue can be examined in a straightforward manner as a periodic Markov jump process. Although there has been research on periodic queues for some decades already, this analysis contains the first explicit solutions of transient and asymptotic distributions of a periodic queue. This demonstrates how useful it is to investigate inhomogeneous Markov jump processes in detail. This chapter provides the non–spatial analogue to chapter 8.

Queues with BMAP input, general service time distribution and infinitely many servers are examined in chapter 5. Again the theory of general Markov jump processes yields the decisive results for a successful analysis. All types of $BMAP/G/\infty$ queues are interpreted as thinned inhomogeneous BMAPs, and thus the transient distributions can be given with results from chapter 3. The intuitively plausible idea of cutting the tail of the service time distribution yields approximation formulae. As an application, bounds for the $BMAP/G/c/c$ loss system are obtained.

Closing the part on classical queueing theory, a new estimation procedure for the parameters of homogeneous BMAPs is introduced in chapter 6. Actually, two procedures are presented. The first is a specification of the EM algorithm to an estimator for BMAPs. However, this approach involves high storage and computation costs. Therefore, a simpler estimation procedure is proposed, and it is shown numerically that this performs almost as well as the EM algorithm.

In part III of this book, the groundwork is lain for the design of spatial queueing models that can serve in the application field of mobile communication networks. In chapter 7, the class of Spatial Markovian Arrival Processes (SMAPs) is defined. This class will provide the arrrival processes for all queues analyzed in this work. SMAPs are a generalization of BMAPs which creates the possibility of modelling spatially distributed batch arrivals, time–inhomogeneous (e.g. periodic) arrival rates, and a general phase space. Although this class of arrival processes is much more general than the class of BMAPs, it retains the same crucial properties that make their analysis tractable. This is shown by the derivation of the most important properties of SMAPs. At the end of the chapter, some examples for SMAPs are given.

Then some practically relevant queues, which are fed by SMAP inputs, are analyzed. These queues feature the definition of spatially variable service time distributions, i.e. the service time of a user may depend on its location at the time instant of arrival. Furthermore, their analysis yields the spatial distribution of users in the system in the transient as well as in the stationary case. Where applicable, loss formulae are given. A further new feature of the queueing models presented in this work is the possibility of including time–inhomogeneous arrival intensities (e.g. periodic ones) and deriving the resulting transient and asymptotic distributions for the respective queues. Since in mobile communication networks there are no waiting users (if the line is busy, a user would give up or try again later), only spatial queues with either infinitely many servers or zero waiting room capacity are examined.

Queues with SMAP arrival processes will be called spatial queues. The class of Markovian spatial queues (i.e. those queues which are Markov processes) is treated in chapter 8. For ease of notational simplicity, this case is examined not in the most general setting, but in terms of the $SMAP/M_t/c/c$ queue. Yet the same method of analysis applies to the $SMAP/M_t/c/c$ queue with spatially variable service time distribution or the $SMAP/PH/c/c$ queue in a completely analogous way. As these queues are loss systems, a formula for the loss probability is derived, too. All these queues can be interpreted as simple models for the cells of a wireless communication network.

Chapter 9 covers the class of spatial infinite server queues. The method of analysis applied to this class allows a variety of features of mobile communication networks to be included in the model. The service time distribution may be general and spatially variable. Furthermore, the arrival rates may depend on time and even stochastic user movements can be modeled in a natural way. In order to make computations more efficient, approximation formulae along with an estimation of the approximation error are given for the most important distributions. As an application in the field of mobile communication networks, it is shown how the analysis of this type of queue can be used in order to determine capacity demands for an optimal positioning of base stations.

In order to practically use stochastic models, it is necessary to obtain statistically validated estimates of the model parameters. In chapter 10, a routine for parameter estimation is introduced for SMAPs. Although crucial for purposes of practical application, the problem of finding some statistically validated parameter estimation is far from trivial. The routine introduced here reduces the complexity of estimation by first neglecting the correlations between inter–arrival times and then re–introducing them again in a parametric form.

I

A CLOSER LOOK INTO THEORY

Chapter 1

MARKOV JUMP PROCESSES

The importance of Markov jump processes for queueing theory is obvious. On the one hand, in stochastic modelling the use of Markov processes makes an analysis of the models much easier. On the other hand, for the number of users in a queueing system a mathematical model must be a jump process. The combination of these, i.e. a jump process with Markov property, will be defined and analyzed as a Markov jump process in the present chapter.

The results that will be obtained in this chapter are important for at least three further concepts in this book. First, they will be used in chapter 2 for the definition and analysis of a broad class of arrival processes for queueing systems. Second, the analysis of Markovian queues as in chapters 4 and 8 requires the derivation of explicit expressions for the distributions of Markov jump processes. Third, even for non–Markovian queues such as the ones analyzed in chapters 5 and 9 the theory of (inhomogeneous) Markov jump processes helps to derive some performance measures.

The following presentation differs from most textbooks in some respects which will be important for the rest of the book. Most popular textbooks analyze Markov jump processes only for the homogeneous case (see Breiman [27], Feller [59] or Doob [52]) and furthermore often for countable state spaces only (see Karlin and Taylor [75]).

In the case of general Markov processes, a restriction to homogeneous processes is no loss of generality, since the state space can always be enlarged to include the time axis. But for Markov jump processes, this construction would destroy the pure jump nature of the sample paths. Thus for Markov jump processes it is useful to regard the inhomogeneous case, too. Further, a restriction to countable state spaces would restrict our possibilities for definitions of spatial arrival processes in chapter 7. Since the presentation for general state spaces is only notationally more complicated than for countable state spaces, we have chosen the more general way.

A nice exception in the textbook presentations are the works by Gikhman and Skorokhod [62, 61]. There we find an analysis for inhomogeneous Markov jump processes with general state spaces. The presentation in this chapter goes along the line of the treatment by Gikhman and Skorokhod. However, the definition and construction are more concise and hopefully less complicated. Further, several useful special cases such as periodic Markov jump processes are analyzed, with results leading to new methods for the analysis of inhomogeneous queues.

This chapter is organized as follows: First, the concept of Markov jump processes will be introduced in section 1. The subsequent section gives a derivation of a simpler form for the transition kernels via the Kolmogorov differential equations. Section 3 provides simplifications for the homogeneous case and an approximation of general Markov jump processes by piecewise homogeneous ones. Finally, the last two sections contain an analysis of two special cases of non–homogeneous Markov jump processes which allow some simplification, too.

1. Definition and Construction

In this section, Markov jump processes will be defined as Markov processes on the space of step functions, which satisfy a technical regularity condition, namely the stochastic continuity. It will be shown that a Markov jump process uniquely determines a generator function, and its distribution will be given in terms of this. Further it will be shown that in the other direction every bounded generator function determines a Markov jump process in distribution.

First, let $(E, \mathcal{E})$ denote some measurable metric space which shall serve as the state space of a Markov jump process. It shall satisfy the measurability of singletons, i.e $\{x\} \in \mathcal{E}$ for every $x \in E$. Denote the space of all step functions $f : I\!R_0^+ \to E$ without explosions by $\mathcal{T}$. As a convention, we will assume that all step functions are right continuous. The variable of the number of users in a queueing system (as a function of time) will behave like a step function. Therefore the stochastic processes that serve as our most simple models for queueing systems shall be supported by the space $\mathcal{T}$ exclusively.

For this chapter, we always assume $\Omega = \mathcal{T}$, i.e. the probability space shall coincide with the space of step functions from time into the space E. Thus every element $\omega \in \Omega$ corresponds to an actual realization conforming with our intuition, and we do not need to modify processes in order to obtain a cadlag path form, as e.g. in Gikhman and Skorokhod [62], chapter 4.

Remark 1.1 Every function $f \in \mathcal{T}$ can be represented as a sequence $z = (z_n = (t_n, x_n) : n \in I\!N_0)$ in a unique way as follows. Let $t_0 := 0$ and

$x_0 := f(0)$ be the start value of f. Having determined $t_n < \infty$, define

$$t_{n+1} := \inf\{t > t_n : f(t) \neq f(t_n)\}$$

as the time instant of the next jump, with $\inf \emptyset := \infty$. If $t_{n+1} = \infty$, then set $x_{n+k} := x_n$ and $t_{n+k} := \infty$ for all $k \in I\!N$. This corresponds to the event that the function f remains in state x_n forever after time t_n. If $t_{n+1} < \infty$, then repeat the iteration to determine $(z_{n+k} : k \geq 2)$. Because of the right continuity of $f \in \mathcal{T}$, we have $f(t_{n+1}) \neq f(t_n)$ if $t_{n+1} < \infty$.

Throughout this chapter, we will keep the notation $\mathcal{F}$ to indicate a fixed σ–algebra on $\Omega = \mathcal{T}$. If P is a probability measure on the measurable space $(\Omega, \mathcal{F})$, then the probability space $X = (\Omega, \mathcal{F}, P)$ is called a **pure jump process**.

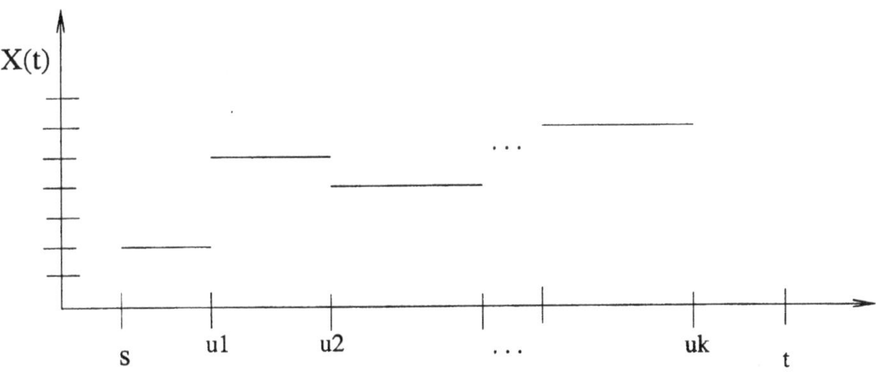

Typical path of a pure jump process

Let pr_t denote the projection to the time instant $t \in I\!R_0^+$, i.e. define $pr_t(f) := f(t)$ for all $f \in \Omega$. We will deal in the following only with canonical filtrations, which will be denoted by $\mathcal{F}_t^s := \sigma(pr_u^{-1} : u \in [s,t])$ and $\mathcal{F}_t := \mathcal{F}_t^0$ for all $s < t \in I\!R_0^+$. Denoting by X_t the random variable on E with distribution $P(pr_t)$, we can write $\mathcal{F}_t^s = \sigma(X_u : s \leq u \leq t)$ for all $s < t \in I\!R_0^+$.

A stochastic process X shall be called a **Markov process** if

$$P(X_t \in A | X_u : 0 \leq u \leq s) = P(X_t \in A | X_s)$$

holds for all $s < t \in I\!R_0^+$ and $A \in \mathcal{E}$. The Markov property (and later the strong Markov property, see below) allows a very thorough understanding of a stochastic process, and thus it is desirable to model queueing systems by Markov processes. The Markov property can be described more lucidly as the property that at any time and in any state, the strict past of the process does not influence the strict future of it. Thus, in order to predict a Markov process, it suffices to know the present state of it.

An $\mathbb{R}_0^+$–valued random variable τ on the probability space $(\Omega, \mathcal{F}, P)$ shall be called a stopping time if $\{\tau \leq t\} \in \mathcal{F}_t$ for all $t \geq 0$. Generalizing the Markov property toward stopping times yields the **strong Markov property**. This is given if

$$P(X_{\tau+t} \in A | X_u : 0 \leq u \leq \tau) = P(X_{\tau+t} \in A | X_\tau)$$

holds for all $A \in \mathcal{E}$, $t \in \mathbb{R}_0^+$ and all stopping times τ.

Remark 1.2 Using exactly the same arguments as in Breiman [27], proposition 15.25, one can show that every Markovian pure jump process is strong Markov.

A more technical property, which shall be exploited in the proof of theorem 1.3, is called stochastic continuity and shall be defined as follows: A Markovian pure jump process is called **stochastically continuous** if

$$\lim_{h \downarrow 0} P(X_{t+h} \in A | X_t = x) = 1_A(x)$$

for all $x \in E$, $A \in \mathcal{E}$ and $t \in \mathbb{R}_0^+$. This notion of continuity states that the marginal distributions $P(X_t)$ as a function of t do not change abruptly. A stochastically continuous Markovian pure jump process is called **Markov jump process** (abbreviated: MJP).

Let X denote an MJP. The first information on X to be gained comes from an examination of the **holding times** of X, which are defined by

$$T(s) := \inf\{t > s : X(t) \neq X(s)\}$$

for all $s \in \mathbb{R}_0^+$, again defining $\inf \emptyset := \infty$. Because of the right continuity of the sample paths, these holding times are stopping times and $X(T(s)) \neq X(s)$ if $T(s) < \infty$.

Theorem 1.3 *For a Markov jump process with holding times $T(s)$ there exist rate functions $u \to q(x, u)$ such that*

$$P(T(s) > t | X(s) = x) = \exp\left(-\int_s^t q(x, u) \, du\right) \tag{1.1}$$

for all $t > s$ and $x \in E$.

Proof: As shown in Gikhman and Skorokhod [61], p.188, the function $t \to P(T(s) > t | X(s) = x)$ is continuous and monotonically non–increasing in

$t > s$ for all $x \in E$ and $s \in I\!R_0^+$. For this result, note that an MJP is stochastically continuous by definition.

Now we can define the measures $q_{s,x}$ on the Borel sets $\mathcal{B}(s,\infty)$ of the half line $]s,\infty[$ by setting $q_{s,x}(]u,v]) := P(T(s) > u|X(s) = x) - P(T(s) > v|X(s) = x)$ for all $s \le u < v$, since these sets generate $\mathcal{B}(s,\infty)$. These measures are Lebesgue–dominated, since the function $t \to P(T(s) > t|X(s) = x)$ is continuous. Hence there is a Radon-Nikodym density $t \to q_s(x,t)$ of any measure $q_{s,x}$ with respect to the Lebesgue measure, i.e. we have $q_{s,x}(K) = \int_K q_s(x,u)du$ for all $K \in \mathcal{B}(s,\infty)$. Specifying $\Psi_{s,x}(t) := q_{s,x}(]t,\infty[)$ yields a set of survivor functions (cf. Davis [49], p.36f) with hazard rates $h_s(x,u) = q_s(x,u)/\Psi_{s,x}(u)$ for all $u > s$, which engenders the representation

$$P(T(s) > t|X(s) = x) = \exp\left(-\int_s^t h_s(x,u)\,du\right) \tag{1.2}$$

for all $t > s$ and $x \in E$, since $P(T(s) > t|X(s) = x) = \Psi_{s,x}(t)$. The Markov property and the equality

$$\{T(s) > t, X(s) = x\} = \{\forall s \le u \le t : X(u) = X(s) = x\}$$

yields the observation

$$\begin{aligned}
\exp\left(-\int_s^t h_s(x,u)\,du\right)& \\
&= P(T(s) > t|X(s) = x) \\
&= P(T(s) > u|X(s) = x) \cdot P(T(u) > t|X(u) = x) \\
&= \exp\left(-\int_s^u h_s(x,v)\,dv\right) \cdot \exp\left(-\int_u^t h_u(x,v)\,dv\right)
\end{aligned}$$

for all $s < u < t$, from which we obtain $h_s(x,t) = h_u(x,t)$ for Lebesgue–almost all t. This means that we can choose one version $q(x,t) := h_s(x,t)$ which is almost everywhere independent of $s < t$, and the statement follows from equation (1.2).
☺

Remark 1.4 The values $q(x,u)$ can be given an interpretation as the infinitesimal rate of a jump from state x at time u. The dependence on u vanishes for homogeneous MJPs.

Remark 1.5 The following example shows that the condition of stochastic continuity is necessary for the statement of theorem 1.3: Let $E := \{0,1\}$, and define f by $f(t) := 0$ for $0 \le t < 1$ and $f(t) := 1$ for $t \ge 1$. Define $P := \delta_f$

to be the Dirac measure on f, i.e. $P(A) := 1$ if $f \in A$ and $P(A) := 0$ if $f \notin A$, for all $A \in \mathcal{F}$. Clearly, $D := (\Omega, \mathcal{F}, P)$ is a pure jump process.

Furthermore, D is a Markov process, since for all $s < t$

$$P(X_t = 1 | X_u : 0 \le u \le s) = \begin{cases} 1, & t \ge 1 \\ 0, & t < 1 \end{cases}$$
$$= P(X_t = 1) = P(X_t = 1 | X_s)$$

holds. However, representation (1.1) does not hold, since there is no function $u \to q(u)$ such that $e^{-\int_0^t q(u)\,du} = 1$ for all $t \ge 1$ and $e^{-\int_0^t q(u)\,du} = 0$ for all $t < 1$.

Next we look not only on the time of the next jump, but also at the state which will be attained immediately after the jump. Here we obtain the following result:

Theorem 1.6 *Let X denote an MJP. For all times $t \in I\!R_0^+$, there are probability distributions Π_t such that*

$$P(u < T(s) \le v, X(T(s)) \in A | X(s) = x)$$
$$= e^{-\int_s^u q(x,w)dw} \cdot \int_u^v e^{-\int_u^t q(x,w)dw} q(x,t) \Pi_t(x, A)\, dt$$

for all $s \le u < v$, $x \in E$ and $A \in \mathcal{E}$.

Proof: For a representation

$$P(u < T(s) \le v, X(T(s)) \in A | X(s) = x) = \int_u^v \Pi_t(x, A)\, dq_{s,x}(t)$$

see Gikhman and Skorokhod [61], p.189f. The rest follows from the particular form $q_{s,x}(]t, \infty[) = e^{-\int_s^t q(x,u)\,du}$ which was derived in the previous theorem. ☺

Remark 1.7 The interpretation for the values $\Pi_t(x, A)$ is that of the following conditional probability: Given that a jump occurs at time t from a state x, this jump is into the state set A.

Now we can define the so–called **infinitesimal transition rates**

$$Q_t(x, A) := q(x,t)\,(\Pi_t(x, A) - 1_A(x))$$

for all $t \in I\!R_0^+$, $A \in \mathcal{E}$ and $x \in E$, with 1_A denoting the characteristic function of the set A. From the representation of theorem 1.6 it is clear that the function

$t \to Q_t(x, A)$ is Lebesgue measurable for all $x \in E$ and $A \in \mathcal{E}$. Furthermore, it can be observed that

$$Q_t(x, A) \begin{cases} \leq 0, & x \in A \\ \geq 0, & x \notin A \end{cases}$$

and $Q_t(x, E) = 0$ for all $x \in E$ and $t \in I\!R_0^+$. A kernel satisfying these conditions shall be called **generator**.

Up to now we have shown that every MJP Lebesgue–uniquely determines an element–wise Lebesgue measurable function $Q : t \to Q_t$ from time to the space of generators on E. This function shall be called the **generator function** of X.

Next we will see that this generator function uniquely determines the distribution of the process X. Define the transition probabilities $P_{st}(x, A) := P(X(t) \in A | X(s) = x)$ and let $P_{st} := (P_{st}(x, A))_{x \in E, A \in \mathcal{E}}$ denote the transition probability kernel (or shortly: transition kernel) from time s to some later time $t > s$. In order to determine these, we will focus on the jump times of the process, which are given as follows: For any time $s \in I\!R_0^+$ at which the observation starts, define $\tau_0(s) := s$ and iteratively

$$\tau_n(s) := \inf\{u > \tau_{n-1}(s) | X(u) \neq X(\tau_{n-1}(s))\}$$

for all $n \in I\!N$. Clearly, $(\tau_n(s) : n \in I\!N_0)$ is a sequence of stopping times for X. The transition probabilities from one jump time to the next are given by theorem 1.6. Thus we have for all $s < t$

$$P_{st}(x, A) = \sum_{n=0}^{\infty} P(X(t) \in A, \tau_n(s) < t < \tau_{n+1}(s) | X(s) = x) \tag{1.3}$$

$$=: \sum_{n=0}^{\infty} P_{st}^{(n)}(x, A) \tag{1.4}$$

with $P_{st}^{(0)}(x, A) = e^{-\int_s^t q(x,w)dw} 1_A(x)$ and iteratively

$$P_{st}^{(n+1)}(x, A) = \int_s^t e^{-\int_s^u q(x,w)dw} \left(Q_u P_{ut}^{(n)}\right)(x, A)\, du$$

$$= \int_s^t e^{-\int_s^u q(x,w)dw} q(x, u) \left(\Pi_u P_{ut}^{(n)}\right)(x, A)\, dt$$

for all $n \in I\!N_0$, recalling that $Q_t(x, A) = q(x, t)\Pi_t(x, A)$ for all $x \in E$ and $A \in \mathcal{E}$. Note that $Q_u P_{ut}^{(n)}$ as well as $\Pi_u P_{ut}^{(n)}$ represent kernel products.

Remark 1.8 The kernels $P_{st}^{(n)}$ can be interpreted as the sub–stochastic transition kernels restricted to all paths which have exactly n jumps in the interval $]s,t]$.

The transition kernels $(P_{st} : s < t \in \mathbb{R}_0^+)$ and any initial distribution determine the set of finite–dimensional marginal distributions of the process and thus according to Kolmogorov's extension theorem (see e.g. Gikhman and Skorokhod [62], pp.108,142) the distribution of the process. Hence the generator function uniquely determines the distribution of the process.

Remark 1.9 The transition probabilities can be computed iteratively by starting with

$$P_0(s,t;x,A) := e^{-\int_s^t q(x,w)dw} 1_A(x)$$

and iterating

$$P_{n+1}(s,t;x,A) := \int_s^t e^{-\int_s^u q(x,w)dw} \int_E Q_u(x,dz) P_n(u,t;z,A) du + e^{-\int_s^t q(x,w)dw} 1_A(x)$$

for all $n \in \mathbb{N}_0$ and $s < t \in \mathbb{R}_0^+$. In the limit, this leads to $P_{st}(x,A) = \lim_{n\to\infty} P_n(s,t;x,A)$ for all $x \in E$ and $A \in \mathcal{E}$.

Remark 1.10 In this iteration, the kernels $P_n(s,t)$ can be interpreted as the sub–stochastic transition kernels restricted to all paths which have at most n jumps in the interval $]s,t]$.

Now let us start from the other end. Assume that a bounded and (element–wise) Lebesgue measurable generator function $Q : t \to Q_t$ is given. We will show that this uniquely determines an MJP.

First, define $q(x,t) := -Q_t(x,\{x\})$ and

$$\Pi_t(x,A) := \begin{cases} Q_t(x, A \setminus \{x\})/q(x,t), & q(x,t) > 0 \\ 1_A(x), & q(x,t) = 0 \end{cases}$$

for all $t \in \mathbb{R}_0^+$. Because of $Q_t(x,E) = 0$, which means that $q(x,t) = Q_t(x, E \setminus \{x\})$, all $\Pi_t(x,.)$ are probability distributions.

Now define a discrete time Markov chain $Y = (\Omega_Y, \mathcal{A}, P^Y)$ with state space $\mathbb{R}_0^+ \times E$ by some initial state $Y_0 = (0,z)$ and the transition probability kernel

$$Q((s,x),(]u,v],A)) := e^{-\int_s^u q(x,w)dw} \cdot \int_u^v e^{-\int_u^t q(x,w)dw} q(x,t)\Pi_t(x,A)\, dt \tag{1.5}$$

for all $s \leq u < v \in I\!R_0^+$, $A \in \mathcal{E}$ and $x \in E$.

Denote $Y = (Y_n = (\tau_n, y_n) : n \in I\!N_0)$. The boundedness of Q yields some $M < \infty$ with $q(x,t) < M$ for all $x \in E$ and $t \in I\!R_0^+$. Denoting by $T = (T_n : n \in I\!N)$ a sequence of iid random variables which are distributed exponentially with parameter M, we obtain

$$P^Y\left(\lim_{n\to\infty} \tau_n < \infty\right) \leq P^T\left(\sum_{n=1}^{\infty} T_n < \infty\right) = 0$$

i.e. the probability for an explosion is zero.

Hence for almost every $\omega \in \Omega_Y$, the sequence $(Y_n(\omega) : n \in I\!N_0)$ determines a step function $f : I\!R_0^+ \to E$ in the way described in remark 1.1. Via this mapping, we have defined a pure jump process $X = (\mathcal{T}, \mathcal{F}, P)$ with state space E. It remains to be shown that X is Markovian (recall remark 1.2) and stochastically continuous. However, these properties are obvious in view of definition (1.5).

The distribution of the process X can be determined in the same way as above: Let

$$N_{st} := \max\{n \in I\!N_0 : \tau_n < t\} - \max\{n \in I\!N_0 : \tau_n < s\}$$

denote the random variable of the number of jumps between times s and $t > s$. Conditioning upon N_{st}, we have

$$P_{st}(x, A) = \sum_{n=0}^{\infty} P(X(t) \in A, N_{st} = n | X(s) = x)$$

and obtain the same form for the transition matrices of X (and hence the same distribution of X) as in equation (1.4).

2. Kolmogorov Differential Equations

In practice, the definition of an MJP is constructive by defining the generator function according to statistical estimators (cf. chapter 6). Thus let us assume from now on that an MJP is given by the definition of a bounded and continuous generator function $Q : t \to Q_t$. Then according to Gikhman and Skorokhod [62], pp.362–365, the equation

$$Q_t = \lim_{h\downarrow 0} \frac{P_{t,t+h} - I}{h}$$

holds for all $t \in I\!R_0^+$, with I denoting the identity kernel.

The infinitesimal transition rate $Q_t(x, A)$ can be regarded as the tendency of the process at time $t \in I\!R_0^+$ to jump from a state $x \in E$ to some state $y \in$

$A \in \mathcal{E}$. The greater the infinitesimal transition rate, the higher the probability that the process performs this jump in a small interval of time. For transition rates of ever greater size, this can lead to so–called explosions which means an accumulation point of jump times. The above assumption of boundedness of the generator function excludes explosions and highly irregular behaviour.

In this chapter, the so–called Kolmogorov differential equations (see again Gikhman and Skorokhod [62], p.314–316) will be exploited in order to find a simpler expression for the transition probabilities. They state that the transition probabilities are differentiable in the following sense:

Theorem 1.11 **(Kolmogorov's forward equation)** *For every $s \in I\!R_0^+$, $A \in \mathcal{E}$ and $x \in E$, the transition probability $P_{st}(x, A)$ is differentiable with respect to t, and*

$$\frac{\partial P_{st}(x, A)}{\partial t} = (P_{st}Q_t)(x, A) = \int_E P_{st}(x, dz)Q_t(z, A)$$

for all $t > s$.

These differential equations lead to expressions of the transition probabilities in terms of the infinitesimal transition rates. This can be proven by solving the Kolmogorov differential equations via the method of successive approximations by Picard and Lindelöf. The solutions are stated in the following theorems.

Theorem 1.12 *Be X a Markov jump process with infinitesimal transition rates $Q_t(x, A)$. The transition probability of X being in some state $y \in A$ after time $t \in I\!R^+$ under the condition of having been in state x at time $s < t$ is given as*

$$P_{st}(x, A) = \sum_{n=0}^{\infty} P_{st}^{(n)}(x, A)$$

with $P_{st}^{(0)}(x, A) := 1_A(x)$, and recursively

$$P_{st}^{(n+1)}(x, A) := \int_s^t \int_E P_{su}^{(n)}(x, dz)Q_u(z, A)du$$

for all $n \in I\!N_0$ and $s < t \in I\!R_0^+$.

Proof: By assumption, the infinitesimal rates $Q_t(x, A)$ are continuous in t for all $x \in E$ and $A \in \mathcal{E}$. Induction by n yields that $P_{st}^{(n)}(x, A)$ is differentiable in t for every $n \in I\!N_0$. Now, direct validation shows

$$\begin{aligned}
\frac{\partial}{\partial t}\sum_{n=0}^{\infty} P_{st}^{(n)}(x,A) &= \frac{\partial}{\partial t}\sum_{n=1}^{\infty} P_{st}^{(n)}(x,A) \\
&= \sum_{n=1}^{\infty} \frac{\partial}{\partial t}\int_s^t \int_E P_{su}^{(n-1)}(x,dz)Q_u(z,A)du \\
&= \sum_{n=1}^{\infty} \int_E P_{st}^{(n-1)}(x,dz)Q_t(z,A) \\
&= \int_E \sum_{n=1}^{\infty} P_{st}^{(n-1)}(x,dz)Q_t(z,A)
\end{aligned}$$

i.e. the Kolmogorov forward equation is satisfied. According to Gikhman and Skorokhod [62], p.317 (theorem 3), the solution to the Kolmogorov forward equation is unique.
Ⓤ

By induction, one can prove that

$$P_{st}^{(n)}(x,A) = \underbrace{\int_s^t \int_s^{u_n} \cdots \int_s^{u_2}}_{n\ integrals} (Q_{u_1} \cdots Q_{u_n})(x,A)\, du_1 \ldots du_n \qquad (1.6)$$

with Q_u denoting the generator at time $u \in [s,t[$.

Analogously to remark 1.9, an iteration for computing the transition probabilities is given by starting with $P_0(s,t;x,A) := 1_A(x)$ and iterating by

$$P_{n+1}(s,t;x,A) := \int_s^t \int_E P_n(s,u;x,dz)Q_u(z,A)du + 1_A(x) \qquad (1.7)$$

for all $n \in I\!N_0$ and $s < t \in I\!R_0^+$. In the limit, this leads to $P_{st}(x,A) = \lim_{n\to\infty} P_n(s,t;x,A)$. This iteration reflects the method of successive approximation by Picard and Lindelöf for solving the Kolmogorov forward differential equations. For the special case of a finite state space, this formula reduces to the iteration given in Bellman [19], p.168, or Kamke [74], p.52.

3. The Homogeneous Case and an Approximation

For the case of a homogeneous Markov jump process , the infinitesimal transition rates are constant in time, and we can define $Q := Q_t$ for all $t \in I\!R_0^+$. The matrix Q shall be called the generator of X. Then the transition probability kernel assumes a particularly tractable form:

Theorem 1.13 *Let X denote a homogeneous Markov jump process with infinitesimal transition rates $(Q(x, A) : x \in E, A \in \mathcal{E})$. Then the transition probability kernel P_{st} can be written as*

$$P_{st} = e^{Q(t-s)} := \sum_{k=0}^{\infty} \frac{(t-s)^k}{k!} Q^k$$

for all $s < t$, with Q^k denoting the kth iteration of the kernel Q.

Proof: It suffices to prove for the expression in theorem 1.12 that

$$P_{st}^{(k)} = \frac{(t-s)^k}{k!} Q^k$$

for all $k \in I\!N_0$ and $s < t$. This is done by induction on k. Beginning with the obvious equality $P_{st}^{(0)} = I = \frac{(t-s)^0}{0!} Q^0$, with I denoting the identity kernel, the iteration step of theorem 1.12 leads to

$$\begin{aligned} P_{st}^{(k+1)}(x, A) &= \int_s^t \int_E P_{su}^{(k)}(x, dz) Q_u(z, A)\, du \\ &= \int_s^t \int_E \frac{(u-s)^k}{k!} Q^k(x, dz) Q(z, A)\, du \end{aligned}$$

for all $k \in I\!N_0$ by induction hypothesis and homogeneity of the transition rates. This yields further

$$\begin{aligned} P_{st}^{(k+1)}(x, A) &= \int_s^t \frac{(u-s)^k}{k!} du \int_E Q^k(x, dz) Q(z, A) \\ &= \frac{(t-s)^{k+1}}{(k+1)!} Q^{k+1}(x, A) \end{aligned}$$

which proves the induction step.
☺

In some cases the computation of the transition probabilities of general Markov jump processes can be reduced to computing some combination of transition probability kernels of homogeneous Markov jump processes. This shall be described in the following.

Choose some time interval $]s, t]$ over which to compute the transition kernel. Since the generator function $Q : t \to Q_t$ is assumed to be continuous, there is a step function $\tilde{Q} : t \to \tilde{Q}_t$, with $\tilde{Q}_t$ being a generator for every t, which approximates the function $Q : t \to Q_t$ uniformly in $[s, t]$, i.e.

$$\begin{aligned} \|\tilde{Q} - Q\|_{s,t} &:= \sup_{s \le u \le t} \|\tilde{Q}_u - Q_u\| \\ &= \sup_{s \le u \le t, x \in E, A \in \mathcal{E}} |\tilde{Q}_u(x, A) - Q_u(x, A)| < \varepsilon \end{aligned}$$

for some $\varepsilon > 0$. Then the following approximation holds:

Theorem 1.14 *Let $\tilde{Q} : u \to \tilde{Q}_u$ be a step function such that $||\tilde{Q} - Q||_{s,t} < \varepsilon$ and $\tilde{Q}_u$ is a generator on E for every $u \in [s,t]$. Then the transition probability kernel P_{st} can be approximated by*

$$\tilde{P}_{st} = \prod_{i=1}^{n} e^{\tilde{Q}_i \cdot (s_i - s_{i-1})}$$

for $s_0 := s$, $s_n := t$ and $\tilde{Q}$ being defined by $\tilde{Q}_u = \tilde{Q}_i$ on $u \in]s_{i-1}, s_i]$ for $i \in \{1,\ldots,n\}$. The error of this approximation is bounded by

$$||\tilde{P}_{st} - P_{st}|| < \varepsilon \cdot (t-s)$$

for all $s < t$.

Proof: Write $\tilde{Q}_u = Q_u + \Delta_u$ for all $u \in [s,t]$. Then by assumption $||\Delta_u|| < \varepsilon$ for all $u \in [s,t]$. Now theorem 1.12 and formula (1.6) yield

$$\begin{aligned}
&\tilde{P}_{st} - P_{st} \\
&= \sum_{k=0}^{\infty} \underbrace{\int_s^t \int_s^{u_k} \cdots \int_s^{u_2}}_{k\ integrals} (Q_{u_1} + \Delta_{u_1}) \ldots (Q_{u_k} + \Delta_{u_k})\, du_1 \ldots du_k \\
&\quad - \sum_{k=0}^{\infty} \underbrace{\int_s^t \int_s^{u_k} \cdots \int_s^{u_2}}_{k\ integrals} Q_{u_1} \ldots Q_{u_k}\, du_1 \ldots du_k \\
&= \sum_{k=1}^{\infty} \underbrace{\int_s^t \int_s^{u_k} \cdots \int_s^{u_2}}_{k\ integrals} \tilde{Q}_{u_1} \ldots \tilde{Q}_{u_{k-1}} \Delta_{u_k}\, du_1 \ldots du_k \\
&\quad + \sum_{k=2}^{\infty} \underbrace{\int_s^t \int_s^{u_k} \cdots \int_s^{u_2}}_{k\ integrals} \tilde{Q}_{u_1} \ldots \tilde{Q}_{u_{k-2}} \Delta_{u_{k-1}} Q_{u_k}\, du_1 \ldots du_k \\
&\quad + \sum_{k=3}^{\infty} \underbrace{\int_s^t \int_s^{u_k} \cdots \int_s^{u_2}}_{k\ integrals} \tilde{Q}_{u_1} \ldots \tilde{Q}_{u_{k-3}} \Delta_{u_{k-2}} Q_{u_{k-1}} Q_{u_k}\, du_1 \ldots du_k \\
&\quad + \ldots \\
&= \int_s^t \tilde{P}_{su} \Delta_u \sum_{k=0}^{\infty} \underbrace{\int_u^t \int_u^{u_k} \cdots \int_u^{u_2}}_{k\ integrals} Q_{u_1} \ldots Q_{u_k}\, du_1 \ldots du_k\, du
\end{aligned}$$

for all $s < t$, whence we obtain

$$\tilde{P}_{st} - P_{st} = \int_s^t \tilde{P}_{su} \Delta_u P_{ut} \, du$$

and because of $||\tilde{P}_{su}|| = ||P_{ut}|| = 1$ for all $u \in [s,t]$, we obtain further

$$||\tilde{P}_{st} - P_{st}|| < \varepsilon \cdot (t - s)$$

which proves the approximation bound.
☺

4. Quasi-Commutability of Generators

A special case of Markov jump processes is given if the following equation (1.8) is satisfied for the generator function. This condition states that the differentiation rule for integrals over the generators can be applied as if the generators were commutable. Therefore the condition shall be called quasi-commutability of the generators. As in the homogeneous case, the transition probabilities as well as the transient distribution assume a rather simple form.

Definition 1.15 If the equation

$$\frac{d}{dt} \left(\int_s^t Q_u du \right)^k = k \cdot \left(\int_s^t Q_u du \right)^{k-1} Q_t \tag{1.8}$$

holds for all $k \in I\!N$ and $s < t$, the generators $(Q_t : t \in I\!R_0^+)$ shall be called **quasi-commutable** .

Remark 1.16 A sufficient condition for equation (1.8) is given if for every $s < t \in I\!R_0^+$ the kernels Q_t and $\int_s^t Q_u du$ are commutable. A sufficient condition for this is the special form

$$Q_t = \lambda(t) \cdot Q$$

for the generators Q_t of X, with $\lambda : I\!R_0^+ \to I\!R_0^+$ describing some time–dependent intensity and Q denoting any generator on E.

Theorem 1.17 *If the generators of X are quasi-commutable, then the transition probability kernels P_{st} assume the form*

$$P_{st} = e^{\int_s^t Q_u du} = \sum_{n=0}^{\infty} \frac{1}{n!} \left(\int_s^t Q_u du \right)^n$$

for all $s < t \in I\!R_0^+$.

Proof: For every $k \in \mathbb{N}$ it will be shown that

$$\underbrace{\int_s^t \int_s^{u_k} \cdots \int_s^{u_2}}_{k\ integrals} Q_{u_1} \dots Q_{u_k} du_1 \dots du_k = \frac{\left(\int_s^t Q_u du\right)^k}{k!}$$

is implied by equation (1.8). For $k = 0$ this holds by definition and for $k = 1$ this is obvious. The induction step for $k + 1$ is seen by first applying the induction hypothesis and then equation (1.8) in

$$\underbrace{\int_s^t \int_s^{u_{k+1}} \cdots \int_s^{u_2}}_{k+1\ integrals} Q_{u_1} \dots Q_{u_{k+1}} du_1 \dots du_{k+1}$$

$$= \int_s^t \frac{\left(\int_s^{u_{k+1}} Q_u du\right)^k}{k!} Q(u_{k+1}) du_{k+1}$$

$$= \frac{1}{k!} \int_s^t \frac{1}{k+1} \frac{d}{du_{k+1}} \left(\int_s^{u_{k+1}} Q_u du\right)^{k+1} du_{k+1}$$

$$= \frac{1}{(k+1)!} \left(\int_s^t Q_u du\right)^{k+1}$$

Now the statement follows by theorem 1.12 and equation (1.6).
☺

Example 1.18 Assume that there is a function $\lambda : \mathbb{R}_0^+ \to \mathbb{R}_0^+$ such that $Q_t = \lambda(t) \cdot Q$ for all $t \in \mathbb{R}_0^+$. Then the transition probability kernel from time $s \in \mathbb{R}_0^+$ to time $t > s$ can be simplified to the form

$$P_{st} = e^{\int_s^t \lambda(u) du \cdot Q} = \sum_{k=0}^{\infty} \frac{\left(\int_s^t \lambda(u) du\right)^k}{k!} Q^k$$

for all $s < t \in \mathbb{R}_0^+$.

5. Periodic Markov Jump Processes

If the infinitesimal transition rates of a Markov jump process are periodic, the Markov jump process will be called periodic, too. The transient and asymptotic distributions for this special case can be determined by examining the imbedded homogeneous Markov chain at times which are multiples of the period length.

For the transient distribution one can prove a recursion formula which simplifies its computation by reducing the time range of integration in formula

(1.6) to at most the period length. An asymptotic distribution does exist if the imbedded Markov chain at every beginning of a period is positive recurrent.

Definition 1.19 A Markov jump process X with generator function Q is called **periodic** with **period** $T > 0$ if $Q_s = Q_{s+T}$ for all $s \in [0, T[$.

5.1 Transient distributions

The computation of the transient distribution can be simplified as follows. The periodicity of the generator yields

$$P_{0,nT} = P_{0,(n-1)T}P_{(n-1)T,nT} = P_{0,(n-1)T}P_{0,T} = P_{0,T}^n$$

for all $n \in I\!N$. Define

$$\lfloor t/T \rfloor := \max\{n \in I\!N_0 : nT \leq t\}$$

as the number of period lengths that have passed until time $t \in I\!R^+$. Now the transition kernels $P_{0,t}$ of X are given by

$$P_{0,t} = P_{0,\lfloor t/T \rfloor T}P_{\lfloor t/T \rfloor T,t} = P_{0,T}^{\lfloor t/T \rfloor}P_{0,t-\lfloor t/T \rfloor T}$$

for all $t \in I\!R_0^+$. This expression allows a computation of the transient distribution at any time $t \in I\!R^+$ without needing to integrate over ranges larger than the period T.

Example 1.20 Assume that the generator function is given by

$$Q_t := \left(1 + \sin\left(\frac{2\pi}{T}t\right)\right) Q$$

for all $t \in I\!R_0^+$, with some constant generator Q. Then the transition kernels $P_{0,s}$ can be expressed by

$$P_{0,s} = \exp\left(\left(s + \int_0^s \sin\left(\frac{2\pi}{T}u\right) du\right) \cdot Q\right)$$

for all $s \leq T$, according to example 1.18. Hence the transition kernels $P_{0,t}$ for $t \in I\!R^+$ are given by

$$P_{0,t} = \left(e^{T \cdot Q}\right)^{\lfloor t/T \rfloor} \exp\left(\left(s + \frac{T}{2\pi}\left(1 - \cos\left(\frac{2\pi}{T}s\right)\right)\right) \cdot Q\right)$$

defining $s := t - \lfloor t/T \rfloor \cdot T$.

5.2 Asymptotic distributions

Before examining the asymptotic behaviour of periodic Markov jump processes, one needs to define the term asymptotic distribution for periodic processes. Denote by $\|\pi\|$ the norm of total variation on E for any signed measure π on E. Further, denote the distribution of a process X at time t given its initial distribution π at time 0 by X_t^π. Now we can formulate

Definition 1.21 Let X denote a periodic Markov jump process with period T. A family $(q_s : s \in [0, T[)$ of probability distributions shall be called a **periodic family of asymptotic distributions** if

$$\|X_{nT+s}^\pi - q_s\| \to 0 \text{ as } n \to \infty$$

for all $s \in [0, T[$, independently from the initial distribution π.

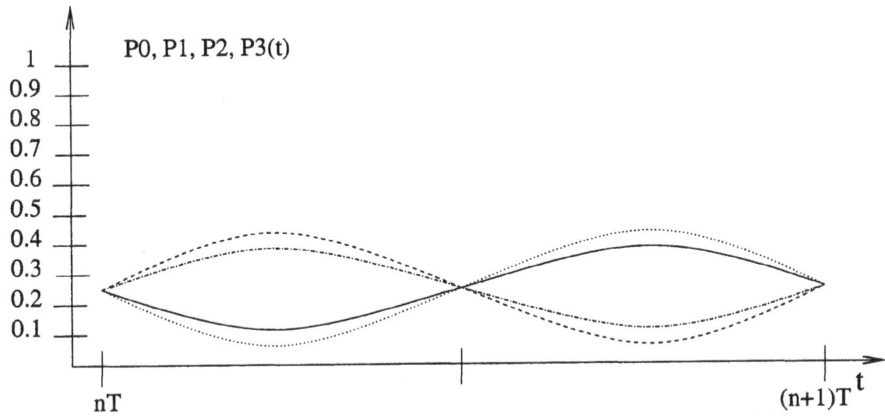

Example for a periodic family $p(t) = (p_0(t), p_1(t), p_2(t), p_3(t))$ of asymptotic distributions (for sufficiently large n)

Theorem 1.22 *If a periodic family* $(q_s : s \in [0, T[)$ *of asymptotic distributions does exist, it is unique.*

Proof: Assume that $(q_s' : s \in [0, T[)$ is another periodic family of asymptotic distributions. Then for every $s \in [0, T[$ and an arbitrary probability measure π

$$\|q_s - q_s'\| \leq \|q_s - X_{nT+s}^\pi\| + \|q_s' - X_{nT+s}^\pi\| \to 0$$

as $n \to \infty$.

Now the main result can be given. It reduces the existence of an asymptotic distribution for the periodic Markov jump process X to the existence of an asymptotic distribution for the embedded Markov chain Y at multiples of the period length. Define $Y = (Y_n : n \in \mathbb{N})$ as the homogeneous Markov chain with transition kernel $P(0,T)$ and let $Y^\pi = (Y_n^\pi : n \in \mathbb{N})$ denote the version of Y with initial distribution π.

Theorem 1.23 *Let X denote a periodic Markov jump process with period T. If X has a periodic family of asymptotic distributions, then Y has a stationary distribution p. If Y has an asymptotic distribution q, then X has a periodic family of asymptotic distributions which is uniquely determined by*

$$q_s = q\ P_{0,s}$$

for all $s \in [0,T[$.

Proof: Let π denote the initial distribution of X at time $t_0 := 0$ and assume that $(q_s : s \in [0,T[)$ is a periodic family of asymptotic distributions. A necessary property of q_0 is

$$q_0 = \lim_{n\to\infty} \pi P_{0,nT} = \left(\lim_{n\to\infty} \pi P_{0,(n-1)T}\right) P_{0,T} = q_0 P_{0,T}$$

which means that $p = q_0$ is a stationary distribution of Y.

Now let q be the asymptotic distribution of Y. Then

$$q_0 = \lim_{n\to\infty} \pi P_{0,nT} = \lim_{n\to\infty} \pi\,(P_{0,T})^n = \lim_{n\to\infty} Y_n^\pi = q$$

does exist.

Using this distribution q, the periodic family of asymptotic distributions $(q_s : s \in [0,T])$ is given by

$$q_s = \lim_{n\to\infty} \pi P_{0,nT} P_{nT,nT+s} = \left(\lim_{n\to\infty} \pi P_{0,nT}\right) P_{0,s} = q P_{0,s}$$

for all $s \in [0,T]$.
☺

Remark 1.24 In order to give conditions for the identity of stationary and asymptotic distribution, we would need to introduce a number of concepts from the theory of Markov chains with general state space. This seems inappropriate for the frame of this book, since the statement in theorem 1.23 is easy to prove and will suffice for further use. The interested reader shall be referred to the monographs by Orey [104], Nummelin [103], as well as Meyn and Tweedie [89].

Example 1.25 We resume example 1.20 and further assume that Q is recurrent and there is a measure q with $qQ = 0$. This implies $qe^{t \cdot Q} = q$ for all $t > 0$ and in particular $qe^{T \cdot Q} = q$. The periodic family of asymptotic distributions $(q_s : s \in [0, T])$ is then given by

$$q_s = q \, \exp\left(\left(s + \frac{T}{2\pi}\left(1 - \cos\left(\frac{2\pi}{T}s\right)\right)\right) \cdot Q\right) = q$$

for all $s \in [0, T[$. This means that in the present special case the periodic family of asymptotic distribution is homogeneous. Hence this special case of periodic MJPs converges asymptotically to a homogeneous MJP. This result is a consequence of the combination of periodicity and quasi–commutability (see Breuer [34] for applications in queueing theory).

Chapter 2

MARKOV–ADDITIVE JUMP PROCESSES

In queueing theory, one field of steady progress is the conception of arrival processes. Back in Erlang's times it was reasonable to assume independent and homogeneous users, which leads to the model of a Poisson arrival process (cf. Doob [52], pp.401ff). However, network traffic has become varied and strongly correlated today. Moreover, limited capacities lead to models which take inhomogeneous traffic into account, in order to profit from day times with less traffic intensity.

The higher correlation of traffic has led to the well–known MAP concept, introduced by Lucantoni et al. [82]. This employs an auxiliary variable, called the phase of the process, which describes stages of varying arrival intensities. It is a notationally simpler form of the processes developed by Neuts [94, 95] in the end of the 1970s. In these the method of phases, which is classical for distributions on $\mathbb{R}^+$ (see Schassberger [118], pp.30ff, and later Neuts [93]), has been exploited for Markov processes. These processes can model correlated arrival streams.

In this chapter, a class of processes shall be introduced, which comprises all Markovian models for arrival streams that have emerged until today and, more importantly, opens some new directions for further progress in the field of arrival processes. One direction is the modelling of traffic which is inhomogeneous in time (as is usual for network traffic). This will be met by the conception of arrival processes that are inhomogeneous in time, too. Further, the growth of mobile communication networks requires models with structured user spaces. This will be met by a generalization of the classical concept of arrivals toward spatial arrival fields.

But first, a short history on the general idea: Inspired by a process developed by Ezhov and Skorokhod [55, 56] in 1969, the term "Markov–additive process" has been coined by Çinlar [41, 42] in 1972. It refers to a two–dimensional

Markov process with transition probabilities that depend on one dimension only. The marginal process in this dimension is a Markov process, too, and shall be called phase process. The marginal process in the other dimension is a process with conditionally independent increments given the phase process.

The first idea of such a process goes back to Neveu [99] in 1961, whose so–called F-process is the class of Markov–additive processes with a finite phase space. A special case of F-processes is the class of BMAPs developed in Lucantoni [80] thirty years later, which are now widely used in queueing theory. Also worth mentioning are the early papers by Ezhov and Skorokhod [55, 56]. The founding work of Çinlar [41, 42] on Markov–additive processes was followed by several studies in the 1970s and 80s, e.g. Arjas, Speed [3], Çinlar [43, 40, 45], or Ney, Nummelin [101, 102] for discrete time.

However, all these studies examined Markov–additive processes in the most general case. In the present paper, the focus is on Markov–additive processes which belong to the class of Markov jump processes. Thus it is possible to explore many additional properties which cannot be derived if the pure jump nature of the sample paths is not exploited. This chapter mainly goes along the contribution by Pacheco and Prabhu [105], which was written after the BMAP concept proved successful as a versatile arrival process for queues. This concept will be generalized in the present chapter and, most importantly, the inhomogeneous case will be analyzed, too.

In the first section of this chapter, Markov–additive jump processes are defined and the transition probabilities are derived in terms of the infinitesimal transition rates. The next section states some elementary properties, mostly resulting from the definition immediately. Under the assumption that the additive part of the state space is a real vector space, Fourier transforms and expectations are derived in section 3. The more specific assumption that the increments for the additive part be non-negative integers in every component of the vector space leads to the same derivations as in section 3 via the easier notion of z-transforms. These processes will be called discrete Markovian arrival processes and analyzed in section 4. In the last section of this chapter, laws of large numbers are given for the special cases of periodic or homogeneous Markov–additive jump processes on real vector spaces.

1. Definition and Transition Probabilities

Markov–additive jump processes will be defined as two–dimensional MJPs which satisfy the condition that the transition probabilities depend on the second dimension only. The first dimension is an additive semi-group and the marginal process on it turns out to have conditionally independent increments given the marginal process in the second dimension.

Let $(N, J) = ((N_t, J_t) : t \in \mathbb{R}_0^+)$ be a two–dimensional Markov jump process with state space $E := \Sigma \times \Phi$. Let $\mathcal{\Sigma}$ and $\mathit{\Phi}$ be σ-algebras on Σ and Φ, respectively, which satisfy $\{x\} \in \mathit{\Sigma}$ and $\{y\} \in \mathit{\Phi}$ for all $x \in \Sigma$ and $y \in \Phi$. Denote $\mathcal{E} := \mathit{\Sigma} \otimes \mathit{\Phi}$ as the product σ-algebra of $\mathit{\Sigma}$ and $\mathit{\Phi}$. Further, let $(\Sigma, +)$ be a semi-group with neutral element $0 \in \Sigma$. Assume that for all $x \in \Sigma$ and $A \in \mathit{\Sigma}$, we also have $A - x := \{s \in \Sigma : s + x \in A\} \in \mathit{\Sigma}$. Then (N, J) is called **Markov–additive jump process** (or shortly: MAJP) if the transition probabilities satisfy the condition

$$P_{st}((x, y), A \times B) = P_{st}((0, y), (A - x) \times B) \tag{2.1}$$

for all $s < t \in I\!R_0^+$, $(x, y) \in E$ and $A \times B \in \mathcal{E}$. Define

$$P_{st}(y, A \times B) := P_{st}((0, y), A \times B)$$

for all $s < t \in I\!R_0^+$, $y \in \Phi$ and $A \times B \in \mathcal{E}$.

Remark 2.1 The above definition postulates that the state space $E := \Sigma \times \Phi$ of a Markov–additive jump process be locally compact, separable and metric. Standard results in topology (cf. Herrlich [67], p.224,118,117) yield that S satisfies this condition if and only if the state spaces Σ and Φ of the marginal processes do so.

Because of equation (2.1), an MAJP is uniquely determined by the probabilities $P_{st}(y, A \times B)$. Since it is a Markov jump process, we can define infinitesimal transition rates

$$Q_t((x, y), A \times B) = \lim_{h \to 0} \frac{P_{t,t+h}((x, y), A \times B) - 1_{A \times B}(x, y)}{h}$$

and the equality

$$Q_t((x, y), A \times B) = Q_t((0, y), (A - x) \times B) \tag{2.2}$$

follows from equation (2.1). Hence, we can define

$$q_t(y, A \times B) := Q_t((0, y), A \times B)$$

for all $t \in I\!R_0^+$, $y \in \Phi$ and $A \times B \in \mathcal{E}$, and further

$$\gamma_t(y) := -q_t(y, \{(0, y)\})$$

and

$$\gamma_t(y, A \times B) := q_t(y, (A \times B) \setminus \{(0, y)\})$$

as well as

$$p_t(y, A \times B) := \begin{cases} \gamma_t(y, A \times B)/\gamma_t(y), & \gamma_t(y) > 0 \\ 1_{A \times B}(0, y), & \gamma_t(y) = 0 \end{cases}$$

Finally, define the kernel Q_t on $\Sigma \times \Phi$ by

$$Q_t((x,y), A \times B) := q_t(y, (A - x) \times B)$$

for all $t \in \mathbb{R}_0^+$, $(x,y) \in E$ and $A \times B \in \mathcal{E}$. The kernel Q_t is called the generator of (N, J) at time t.

Remark 2.2 By definition, a Markov–additive jump process is translation invariant or homogeneous in the first component. This leads to a self–similar structure of the generators $(Q_t : t \in \mathbb{R}_0^+)$, which can be illustrated in the case of a homogeneous Markov–additive jump process with Φ being finite and $\Sigma = \mathbb{N}_0$ (the so–called Batch Markovian Arrival Process or BMAP, see chapter 3). Here, the generator Q, which is constant in time, takes the form

$$Q = \begin{pmatrix} D_0 & D_1 & D_2 & D_3 & \cdots \\ & D_0 & D_1 & D_2 & \cdots \\ & & D_0 & D_1 & \ddots \\ & & & D_0 & \ddots \\ & & & & \ddots \end{pmatrix}$$

with $m \times m$-matrices $(D_n : n \in \mathbb{N}_0)$, m being the size of Φ. Omitting the first row and first column, one obtains the matrix Q again. This self–similarity yields many simplifications for the analysis of queues with BMAP arrivals. For a so–called level–dependent BMAP, which does not possess this self–similarity, and for the analysis of queues with level–dependent BMAP arrivals see Hofmann [69, 70]. More on BMAPs will be explicated in chapter 3.

As in chapter 1, the transition probabilities of the process will be derived as the solutions of the Kolmogorov differential equations via the method of successive approximations by Picard and Lindelöf. Thus theorem 1.11 and equations (2.1) and (2.2) lead to the following forms for the Kolmogorov differential equations:

$$\frac{\partial P_{s,t}(y, A \times B)}{\partial t} = \int_{\Sigma \times \Phi} q_t(w, (A - v) \times B) \, P_{s,t}(y, d(v,w))$$

for all $t > s$ (**Kolmogorov's forward equation**) and

$$\frac{\partial P_{s,t}(y, A \times B)}{\partial s} = -\int_{\Sigma \times \Phi} P_{s,t}(w, (A - v) \times B) \, q_s(y, d(v,w))$$

for all $s < t$ (**Kolmogorov's backward equation**).

Both differential equations contain a convolution in the first dimension. This is a consequence of the additivity which is defined on the marginal state space

Σ. The convolution form will be preserved in the transition probabilities, as the next results show. First, remark 1.9 yields together with equations (2.1) and (2.2) that the transition probabilities can be computed iteratively by starting with

$$P_0(s,t;y,A\times B) := e^{-\int_s^t \gamma_\tau(y)d\tau} 1_{A\times B}(0,y)$$

and iterating by

$$P_{n+1}(s,t;y,A\times B) \\ = \int_s^t \int_S P_n(u,t;w,(A-v)\times B) e^{-\int_s^u \gamma_\tau(y)d\tau} \gamma_u(y,d(v,w))du \\ + e^{-\int_s^t \gamma_\tau(y)d\tau} 1_{A\times B}(0,y)$$

for all $n \in \mathbb{N}_0$ and $s < t \in \mathbb{R}_0^+$. Then we have $P_{s,t} = \lim_{n\to\infty} P_n(s,t)$ for all $s < t \in \mathbb{R}_0^+$. For an interpretation of the kernels $P_n(s,t)$, remark 1.10 remains valid.

As an immediate consequence from theorem 1.12 and equations (2.1) and (2.2), we obtain a representation of the transition probabilities of an MAJP in the form

$$P_{s,t}(y,A\times B) = \sum_{n=0}^{\infty} P_{s,t}^{(n)}(y,A\times B)$$

with

$$P_{s,t}^{(0)}(y,A\times B) := 1_{A\times B}(0,y) \tag{2.3}$$

and recursively

$$P_{s,t}^{(n+1)}(y,A\times B) := \int_s^t \int_{\Sigma\times\Phi} P_{s,u}^{(n)}(y,d(v,w)) q_u(w,(A-v)\times B)du \tag{2.4}$$

for all $n \in \mathbb{N}_0$. Formula (1.6) is specified to

$$P_{s,t}^{(n)}(y,A\times B) \\ = \underbrace{\int_s^t \int_s^{u_n} \cdots \int_s^{u_2}}_{n\ integrals} (Q_{u_1}\cdots Q_{u_n})\,((0,y),A\times B)\,du_1\ldots du_n \tag{2.5}$$

with Q_u denoting the generator at time $u \in [s,t]$.

2. Elementary Properties

Some elementary properties for Markov–additive jump processes can be taken from Çinlar [41, 42] or Ezhov and Skorokhod [55, 56], who examined the more general class of Markov–additive processes. This section contains some properties which are immediate consequences of the definition of MAJPs. Obviously, the basic sample path properties of Markov jump processes hold for the subclass of Markov–additive jump processes, too. Furthermore, the distribution of the marginal holding time in a state of the additive space Σ can be given.

Let (N, J) be a Markov–additive jump process with state space $\Sigma \times \Phi$. The definitions of MJPs and MAJPs immediately show that the marginal process J is a Markov jump process with state space Φ and transition probabilities

$$P_{s,t}^{\Phi}(y, B) = P_{s,t}(y, \Sigma \times B) \tag{2.6}$$

for all $s < t \in I\!R_0^+$, $y \in \Phi$ and $B \in \mathit{\Phi}$. This observation gives rise to a method of analysis which regards the marginal process J as the independent underlying Markov jump process that conditions the additive process N (see theorem 2.5). The marginal state space Φ shall be called **phase space** of (N, J), a single element of Φ is called a **phase**. The marginal process J shall be called **phase process** of (N, J), while N shall be called **additive process** of (N, J). The next theorem shows that N has conditionally independent increments given J.

Theorem 2.3 *Let (N, J) be a Markov–additive jump process with state space $\Sigma \times \Phi$. Define $T := \inf\{t \in I\!R^+ : N_t \neq N_0\}$ as the holding time of the marginal process N in a state of Σ. Further, define the sub–stochastic kernel $P(T > t)$ on Φ by*

$$P(T > t)(y, B) := P(T > t, J_t \in B | J_0 = y)$$

for all $t \in I\!R^+$, $y \in \Phi$ and $B \in \mathit{\Phi}$. Then for all $s, t \in I\!R^+$,

$$P(T > t + s) = P(T > t)P(T > s)$$

and furthermore,

$$P(T > t)(y, B) = \sum_{n=0}^{\infty} \underbrace{\int_0^t \int_0^{u_n} \cdots \int_0^{u_2}}_{n\ integrals} \left(Q_{u_1}^{(0)} \cdots Q_{u_n}^{(0)}\right)(y, B)\, du_1 \ldots du_n \tag{2.7}$$

with

$$Q_t^{(0)}(y, B) := Q_t(y, \{0\} \times B)$$

for all $t \in I\!R_0^+$, $y \in \Phi$ and $B \in \mathit{\Phi}$.

Proof: The first equation follows from the definition of $P(T > t)$ and the fact that (N, J) is a Markov–additive jump process. For the second equation,

note that the holding time T is identical to the first hitting time T_A of the set $A := \Sigma \backslash \{0\} \times \Phi$. This equals the life time of the cut–off Markov jump process with state space Φ and generators $(Q_t^{(0)} : t \in I\!R_0^+)$. Then statement (2.7) is merely the form for the transition probability kernel of the cut–off process, which follows from formula (2.5).
☺

Remark 2.4 This result contains the statements of theorems 6.9 and 6.11 in Pacheco and Prabhu [105] as the special case of (N, J) being homogeneous, $\Sigma = I\!N_0^r$ and Φ being countable. This follows from theorem 1.13.

For later use, a fundamental theorem will be stated which asserts the existence of a representation of the infinitesimal transition rates (resp. the transition probabilities) as the product of a kernel on the phase space Φ (which is a function $\Phi \times \Phi \to [0, 1]$) and a kernel from Φ into the additive space Σ (which is a function $\Phi \times \Sigma \to [0, 1]$).

Theorem 2.5 *For every $t \in I\!R_0^+$ and $y \in \Phi$, there is a kernel $K_{t,y} : \Phi \times \Sigma \to I\!R$ such that*

$$q_t(y, A \times B) = \int_B q_t(y, \Sigma \times dz) K_{t,y}(z, A)$$

for all $A \in \Sigma$ and $B \in \Phi$. Furthermore, for every $s < t \in I\!R_0^+$ and $y \in \Phi$, there is a kernel $L_{s,t,y} : \Phi \times \Sigma \to I\!R$ such that

$$P_{s,t}(y, A \times B) = \int_B P_{s,t}(y, \Sigma \times dz) L_{s,t,y}(z, A)$$

for all $A \in \Sigma$ and $B \in \Phi$. The kernels $K_{t,y}$ and $L_{s,t,y}$ are almost surely uniquely determined.

Proof: See Bauer [15], p.397 (with $C = \Sigma \times \Phi$), or Bourbaki [26], p.39 (with $p = pr_2$), since the signed measure $q_t(y, .)$ can be represented as the difference of two finite measures.

3. Markov–additive jump processes on a Real Vector Space

Now assume that $(\Sigma, +)$ is a real vector space. Then $(\Sigma, +)$ has a base $(b_j : j \in I)$ with some index set I, and every element $x \in \Sigma$ has a unique representation $(x_j : j \in I)$ in $I\!R^I$ with respect to this base (cf. Nef [92], p.46). Denote the bijective mapping between an element $x \in \Sigma$ and its representation $(x_j : j \in I)$ by $f : \Sigma \to I\!R^I$. In order to simplify the notation in the following, an element $x \in \Sigma$ and its representation $f(x) \in I\!R^I$ shall be identified. Further, denote the projection from $\mathbb{R}^I$ to the j-th component by $pr_j : \mathbb{R}^I \to \mathbb{R}$.

Under this assumption, transforms as well as expectations can be derived for the marginal process N of a Markov–additive jump processes (N, J). Using the structure of the vector space, we can define the expectation of a Markov–additive jump process with state space $\Sigma \times \Phi$ as follows: Let (N, J) be a Markov–additive jump process with state space $\Sigma \times \Phi$ and assume that $(\Sigma, +)$ is a real vector space. The **expectation kernel** $E(N_t - N_s)$ of the marginal process N during the time interval $]s, t]$ is defined as the kernel on Φ with entries

$$E(N_t - N_s)(y, B) := (E(pr_j(N_t - N_s) \cdot 1_B(J_t)|J_s = y) : j \in I)$$

for all $s < t$, $y \in \Phi$ and $B \in \mathcal{\Phi}$.

For real valued random variables, the Fourier transform has proved very useful, especially for determining moments (cf. Bauer [15], pp.183-223). The same analytical method shall be applied here for every dimension of the vector space Σ. The **Fourier transform kernel** $N^*_{s,t}$ of (N, J) over the time interval $]s, t]$ shall be defined as the function $r \to N^*_{s,t}(r)$ with

$$N^*_{s,t}(r)(y, B) := \left(\int_{-\infty}^{\infty} e^{i \cdot rc}\, P_{st}(y, pr_j^{-1}(dc) \times B) : j \in I\right)$$

for all $r \in \mathbb{R}$, $y \in \Phi$ and $B \in \mathcal{\Phi}$, with i denoting the imaginary unit in the space $\mathbb{C}$ of the complex numbers.

Remark 2.6 Note that by theorem 2.5, the distribution $P_{st}(y, pr_j^{-1}(dc) \times B)$ in $c \in \mathbb{R}$ can be written as

$$P_{st}(y, pr_j^{-1}(dc) \times B) = \int_B P_{st}(y, \Sigma \times dz) L_{s,t,y}(z, dc)$$

with some kernel $L_{s,t,y}$ from Φ to $\mathbb{R}$. Hence the representation

$$pr_j(N^*_{s,t}(r)(y, B)) = \int_B \int_{-\infty}^{\infty} e^{i \cdot rc}\, L_{s,t,y}(z, dc)\; P_{st}(y, \Sigma \times dz)$$

holds and thus every component $pr_j(N^*_{s,t}(r))$ is a kernel on Φ.

Theorem 2.7 *For every $j \in I$, $r \in \mathbb{R}$ and $t \in \mathbb{R}_0^+$, define the kernel $\varphi_{Q_t^{(j)}}(r)$ on Φ by its entries*

$$\begin{aligned}
\varphi_{Q_t^{(j)}}(r)(y, B) &:= \int_{-\infty}^{\infty} e^{i \cdot rc}\, q_t(y, pr_j^{-1}(dc) \times B) \\
&= \int_B \int_{-\infty}^{\infty} e^{i \cdot rc}\, K_{t,y}(z, pr_j^{-1}(dc))\; q_t(y, \Sigma \times dz)
\end{aligned}$$

for all $y \in \Phi$ and $B \in \mathit{\Phi}$, with some kernel $K_{t,y}$ from Φ to Σ (see theorem 2.5 for the second equality). Then the jth component of the Fourier transform kernel of (N, J) over the time interval $]s, t]$ can be written as

$$pr_j(N^*_{s,t}(r)) = \sum_{n=0}^{\infty} \underbrace{\int_s^t \int_s^{u_n} \cdots \int_s^{u_2}}_{n \text{ integrals}} \varphi_{Q^{(j)}_{u_1}}(r) \ldots \varphi_{Q^{(j)}_{u_n}}(r)\, du_1 \ldots du_n$$

for all $r \in I\!R$, with the summand for $n = 0$ being the identity kernel I on Φ.

Proof: Fix $j \in I$ as well as $y \in \Phi$ and $B \in \mathit{\Phi}$. According to formula (2.5), the jth component of $N^*_{s,t}(r)(y, B)$ equals

$$\begin{aligned}
&pr_j(N^*_{s,t}(r)(y, B)) \\
&\quad = \sum_{n=0}^{\infty} \int_{-\infty}^{\infty} e^{i \cdot rc}\, P^{(n)}_{st}(y, pr_j^{-1}(dc) \times B) \\
&\quad = \sum_{n=0}^{\infty} \int_s^t \int_s^{u_n} \cdots \int_s^{u_2} \int_{-\infty}^{\infty} e^{i \cdot rc} \\
&\qquad\qquad (Q_{u_1} \cdots Q_{u_n})\left((0, y), pr_j^{-1}(dc) \times B\right) du_1 \ldots du_n \\
&\quad = \sum_{n=0}^{\infty} \int_s^t \int_s^{u_n} \cdots \int_s^{u_2} \int_{I\!R \times \Phi} q_{u_1}(y, pr_j^{-1}(dc_1) \times dy_1) e^{i \cdot rc_1} \ldots \\
&\qquad\qquad \ldots \int_{I\!R \times \Phi} q_{u_{n-1}}(y_{n-2}, pr_j^{-1}(dc_{n-1}) \times dy_{n-1}) e^{i \cdot rc_{n-1}} \\
&\qquad\qquad \int_{I\!R \times B} q_{u_n}(y_{n-1}, pr_j^{-1}(dc_n) \times dy_n) e^{i \cdot rc_n}\, du_1 \ldots du_n
\end{aligned}$$

by writing out the convolution $c = c_1 + \ldots + c_n$ explicitly. Further, application of theorem 2.5 leads to

$$\begin{aligned}
&pr_j(N^*_{s,t}(r)(y, B)) \\
&= \sum_{n=0}^{\infty} \int_s^t \int_s^{u_n} \cdots \int_s^{u_2} \int_{\Phi} \int_{-\infty}^{\infty} e^{i \cdot rc_1} K_{u_1,y}(y_1, pr_j^{-1}(dc_1)) q_{u_1}(y, \Sigma \times dy_1)) \\
&\ldots \int_{\Phi} \int_{-\infty}^{\infty} e^{i \cdot rc_{n-1}} K_{u_{n-1}, y_{n-2}}(y_{n-1}, pr_j^{-1}(dc_{n-1})) q_{u_{n-1}}(y_{n-2}, \Sigma \times dy_{n-1}) \\
&\int_B \int_{-\infty}^{\infty} e^{i \cdot rc_n} K_{u_n, y_{n-1}}(y_n, pr_j^{-1}(dc_n)) q_{u_n}(y_{n-1}, \Sigma \times dy_n)\, du_1 \ldots du_n
\end{aligned}$$

$$= \sum_{n=0}^{\infty} \int_s^t \int_s^{u_n} \cdots \int_s^{u_2}$$

$$\int_\Phi \varphi_{Q_{u_1}^{(j)}}(r)(y, dy_1) \ldots \int_B \varphi_{Q_{u_n}^{(j)}}(r)(y_{n-1}, dy_n)\, du_1 \ldots du_n$$

$$= \sum_{n=0}^{\infty} \int_s^t \int_s^{u_n} \cdots \int_s^{u_2} \left(\varphi_{Q_{u_1}^{(j)}}(r) \ldots \varphi_{Q_{u_n}^{(j)}}(r)\right)(y, B)\, du_1 \ldots du_n$$

which was to be proven.
☺

Theorem 2.8 *Define the kernel $M_t^{(j)}$ on Φ by*

$$M_t^{(j)}(y, B) := \int_{-\infty}^{\infty} c\, q_t(y, pr_j^{-1}(dc) \times B)$$

for all $t \in I\!R$, $j \in I$, $y \in \Phi$ and $B \in \mathcal{\Phi}$. Then the expectation kernel of the marginal process N over the time interval $]s, t]$ is given by

$$E(N_t - N_s) = \left(\int_s^t P_{su}^\Phi\, M_u^{(j)}\, P_{ut}^\Phi\, du : j \in I\right)$$

Proof: The expectation $E(N_t - N_s)$ can be computed as

$$E(N_t - N_s) = -i \cdot \frac{d}{dr} N_{s,t}^*(r)\bigg|_{r=0}$$

recognizing that the differentiation can be performed component–wise. Then the jth component of $E(N_t - N_s)$ is given by

$$pr_j(E(N_t - N_s)) = -i \cdot \sum_{n=1}^{\infty} \underbrace{\int_s^t \int_s^{u_n} \cdots \int_s^{u_2}}_{n \text{ integrals}}$$

$$\frac{d}{dr}\left(\varphi_{Q_{u_1}^{(j)}}(r) \ldots \varphi_{Q_{u_n}^{(j)}}(r)\right)\bigg|_{r=0} du_1 \ldots du_n$$

according to the preceding theorem 2.7. Applying the product rule of differentiation, this equals

$$pr_j(E(N_t - N_s))$$

$$= -i \cdot \sum_{n=1}^{\infty} \underbrace{\int_s^t \int_s^{u_n} \cdots \int_s^{u_2}}_{n \text{ integrals}} \sum_{l=1}^{n} \varphi_{Q_{u_1}^{(j)}}(0) \ldots \varphi_{Q_{u_{l-1}}^{(j)}}(0) \left(\frac{d}{dr}\varphi_{Q_{u_l}^{(j)}}(r)\bigg|_{r=0}\right)$$

$$\varphi_{Q_{u_{l+1}}^{(j)}}(0) \ldots \varphi_{Q_{u_n}^{(j)}}(0)\, du_1 \ldots du_n$$

$$= \sum_{l=1}^{\infty} \sum_{n=l}^{\infty} \underbrace{\int_s^t \int_s^{u_n} \cdots \int_s^{u_{l+2}}}_{n-l \ integrals} \int_s^{u_{l+1}}$$

$$\underbrace{\int_s^{u_l} \cdots \int_s^{u_2}}_{l-1 \ integrals} \varphi_{Q_{u_1}^{(j)}}(0) \ldots \varphi_{Q_{u_{l-1}}^{(j)}}(0) du_1 \ldots du_{l-1}$$

$$\left(-i \cdot \frac{d}{dr} \varphi_{Q_{u_l}^{(j)}}(r) \bigg|_{r=0} \right) du_l \ \varphi_{Q_{u_{l+1}}^{(j)}}(0) \ldots \varphi_{Q_{u_n}^{(j)}}(0) \ du_{l+1} \ldots du_n$$

Acknowledging that

$$\begin{aligned} -i \cdot \frac{d}{dr} \varphi_{Q_{u_l}^{(j)}}(r) \bigg|_{r=0} (y, B) &= \int_{-\infty}^{\infty} c \ q_{u_l}(y, pr_j^{-1}(dc) \times B) \\ &= M_{u_l}^{(j)}(y, B) \end{aligned}$$

is the expectation kernel of the jth component and

$$\varphi_{Q_u^{(j)}}(0)(y, B) = q_u(y, \Sigma \times B)$$

is the infinitesimal transition rate of the phase process, this leads to

$$\begin{aligned} &pr_j(E(N_t - N_s)) \\ &= \int_s^t \left(\sum_{l=1}^{\infty} \int_s^{u_l} \cdots \int_s^{u_2} \varphi_{Q_{u_1}^{(j)}}(0) \ldots \varphi_{Q_{u_{l-1}}^{(j)}}(0) \ du_1 \ldots du_{l-1} \right) M_{u_l}^{(j)} \\ &\quad \left(\sum_{n=l}^{\infty} \int_{u_l}^t \int_{u_l}^{u_n} \cdots \int_{u_l}^{u_{l+2}} \varphi_{Q_{u_{l+1}}^{(j)}}(0) \ldots \varphi_{Q_{u_n}^{(j)}}(0) \ du_{l+1} \ldots du_n \right) du_l \\ &= \int_s^t P_{su_l}^{\Phi} M_{u_l}^{(j)} P_{u_l t}^{\Phi} \ du_l \end{aligned}$$

which was to be proven.
☺

4. Discrete Markovian Arrival Processes

A Markov–additive process (N, J) on a real vector space (of dimension $|I|$) shall be called **Markovian arrival process** (or shortly: MAP) if the support of the marginal process N is on $(I\!R_0^+)^I$, i.e. if $P(pr_j(N_t) \in I\!R_0^+) = 1$ for all $j \in I$ and $t \in I\!R_0^+$. This reflects the notion that an arrival constitutes a positive addition to some load of a system. Thus a process consisting of arrivals only cannot reach any negative values.

Remark 2.9 The term Markovian arrival process along with its abbreviation MAP has initially been introduced in Lucantoni et al. [82]. Lucantoni's MAP was a natural generalization of the Poisson process, which resulted from the introduction of a phase space. However, the notion of arrivals was confined to single entities without further structure. Markovian arrival processes as introduced in the present book are a natural generalization of Lucantoni's MAPs in the sense that they provide them with structured arrival spaces. In our presentation of spatial arrival processes in chapter 7 it will be necessary to give a versatile structure to arrival spaces, which is the reason for this generalization.

In queueing theory, the most important variable is that of the number of users or customers in the system (which is the queue or the queueing network). The values of this are limited to the set $I\!N_0$ of non-negative integers. Thus for most applications in queueing theory, the use of MAPs with $I\!N_0^I$-valued marginal process N suffices. Such processes shall be called discrete MAPs. Results for this special case will be useful for the definition of spatial MAPs in chapter 7 and for the analysis of spatial queues. For discrete MAPs, the easier notion of a z-transform serves the same purposes as the Fourier transform. In this section, the same analysis as in the preceding section shall be carried out using z-transforms.

The homogeneity of the first component implies that MAPs have only non-negative increments in the first dimension. This follows from property (2.1), since for every $x \in \Sigma$, $z \in I\!R^+$ and any dimension $j \in I$ of the vector space,

$$P_{st}((x,y), pr_j^{-1}(pr_j(x) - z) \times B) = P_{st}(y, pr_j^{-1}(-z) \times B) = 0$$

because of $pr_j^{-1}(-z) \notin (I\!R_0^+)^I$.

Define the **z-transform kernel** of (N, J) over the time interval $]s,t]$ as the function $z \to N_{st}(z)$ with values being the kernels on Φ which are determined by

$$N_{st}(z)(y,B) := \left(\sum_{n=0}^{\infty} P(pr_j(N_t - N_s) = n, J_t \in B | J_s = y) \; z^n : j \in I \right)$$

for all $z \in \mathbb{C}$ with $|z| \le 1$, $y \in \Phi$ and $B \in \mathit{\Phi}$.

Theorem 2.10 *For every $j \in I$, $k \in I\!N_0$ and $t \in I\!R_0^+$, define the kernels $Q_t^{(j)}(k)$ on Φ by*

$$Q_t^{(j)}(k)(y,B) := q_t(y, pr_j^{-1}(k) \times B)$$

for all $y \in \Phi$ and $B \in \varPhi$. Then the jth component of the z-transform of (N, J) over the time interval $]s,t]$ can be written as

$$N_{st}^{(j)}(z) = \sum_{n=0}^{\infty} \underbrace{\int_s^t \int_s^{u_n} \cdots \int_s^{u_2}}_{n \text{ integrals}} \sum_{k=0}^{\infty} Q_{u_1}^{(j)}(k) z^k \cdots \sum_{k=0}^{\infty} Q_{u_n}^{(j)}(k) z^k \, du_1 \ldots du_n$$

for all $z \in \mathbb{C}$ with $|z| \leq 1$.

Remark 2.11 Note that $\sum_{k=0}^{\infty} Q_{u_l}^{(j)}(k) z^k$ is a kernel on Φ for every $l \in \mathbb{N}$ and thus every product $\sum_{k=0}^{\infty} Q_{u_1}^{(j)}(k) z^k \ldots \sum_{k=0}^{\infty} Q_{u_n}^{(j)}(k) z^k$ is a kernel on Φ, too.

Proof: Fix $j \in I$ as well as $y \in \Phi$ and $B \in \varPhi$. According to formula (2.5), the jth component of $N_{st}(z)(y, B)$ equals

$$\begin{aligned}
&pr_j(N_{st}(z)(y, B)) \\
&\quad = \sum_{k=0}^{\infty} \sum_{n=0}^{\infty} P_{st}^{(n)}(y, pr_j^{-1}(k) \times B) \cdot z^k \\
&\quad = \sum_{k=0}^{\infty} \sum_{n=0}^{\infty} \int_s^t \int_s^{u_n} \cdots \int_s^{u_2} (Q_{u_1} \cdots Q_{u_n})\left((0, y), pr_j^{-1}(k) \times B\right) \\
&\qquad\qquad du_1 \ldots du_n \cdot z^k \\
&\quad = \sum_{n=0}^{\infty} \int_s^t \int_s^{u_n} \cdots \int_s^{u_2} \sum_{k=0}^{\infty} \sum_{k_1+\ldots+k_n=k} \left(Q_{u_1}^{(j)}(k_1) \ldots Q_{u_n}^{(j)}(k_n)\right)(y, B) \\
&\qquad\qquad z^k \, du_1 \ldots du_n \\
&\quad = \sum_{n=0}^{\infty} \int_s^t \int_s^{u_n} \cdots \int_s^{u_2} \left(\sum_{k=0}^{\infty} Q_{u_1}^{(j)}(k) z^k \ldots \sum_{k=0}^{\infty} Q_{u_n}^{(j)}(k) z^k\right)(y, B) \\
&\qquad\qquad du_1 \ldots du_n
\end{aligned}$$

since the z-transform of the convolution of the kernels $Q_{u_1}^{(j)}(k_1,) \ldots, Q_{u_n}^{(j)}(k_n)$ equals the product of the z-transforms of $Q_{u_1}^{(j)}(k_1), \ldots, Q_{u_n}^{(j)}(k_n)$.
☺

Theorem 2.12 *The expectation kernel of the marginal process N over the time interval $]s,t]$ is given by*

$$E(N_t - N_s) = \left(\int_s^t P_{su}^{\Phi} \sum_{k=1}^{\infty} k \cdot Q_u^{(j)}(k) \, P_{ut}^{\Phi} \, du : j \in I\right)$$

Proof: The expectation kernel $E(N_t - N_s)$ can be computed as

$$E(N_t - N_s) = \frac{d}{dz} N_{st}(z) \bigg|_{z=1}$$

recognizing that the differentiation can be performed component–wise. Then the jth component of $E(N_t - N_s)$ is given by

$$pr_j(E(N_t - N_s)) = \sum_{n=1}^{\infty} \underbrace{\int_s^t \int_s^{u_n} \cdots \int_s^{u_2}}_{n\ integrals}$$

$$\frac{d}{dz} \left(\sum_{k=0}^{\infty} Q_{u_1}^{(j)}(k) z^k \cdots \sum_{k=0}^{\infty} Q_{u_n}^{(j)}(k) z^k \right) \bigg|_{z=1} du_1 \ldots du_n$$

Denote the generator of the phase process at time u by $Q^{\Phi}(u)$ and note that

$$Q^{\Phi}(u) = \sum_{k=0}^{\infty} Q_u^{(j)}(k)$$

is independent of $j \in I$. Applying the product rule of differentiation yields

$$pr_j(E(N_t - N_s))$$

$$= \sum_{n=1}^{\infty} \int_s^t \int_s^{u_n} \cdots \int_s^{u_2} \sum_{l=1}^{n} Q_{u_1}^{\Phi} \cdots Q_{u_{l-1}}^{\Phi} \left(\sum_{k=1}^{\infty} k Q_{u_l}^{(j)}(k) \right) Q_{u_{l+1}}^{\Phi} \cdots Q_{u_n}^{\Phi} \, du_1 \ldots du_n$$

$$= \sum_{l=1}^{\infty} \sum_{n=l}^{\infty} \underbrace{\int_s^t \int_s^{u_n} \cdots \int_s^{u_{l+2}}}_{n-l\ integrals} \int_s^{u_{l+1}} \underbrace{\int_s^{u_l} \cdots \int_s^{u_2}}_{l-1\ integrals} Q_{u_1}^{\Phi} \cdots Q_{u_{l-1}}^{\Phi} du_1 \ldots du_{l-1}$$

$$\left(\sum_{k=1}^{\infty} k Q_{u_l}^{(j)}(k) \right) du_l Q_{u_{l+1}}^{\Phi} \cdots Q_{u_n}^{\Phi} \, du_{l+1} \ldots du_n$$

$$= \int_s^t \left(\sum_{l=1}^{\infty} \int_s^{u_l} \cdots \int_s^{u_2} Q_{u_1}^{\Phi} \cdots Q_{u_{l-1}}^{\Phi} du_1 \ldots du_{l-1} \right) \sum_{k=1}^{\infty} k Q_{u_l}^{(j)}(k)$$

$$\left(\sum_{n=l}^{\infty} \int_{u_l}^t \int_{u_l}^{u_n} \cdots \int_{u_l}^{u_{l+2}} Q_{u_{l+1}}^{\Phi} \cdots Q_{u_n}^{\Phi} \, du_{l+1} \ldots du_n \right) du_l$$

$$= \int_s^t P_{su_l}^{\Phi} \left(\sum_{k=1}^{\infty} k Q_{u_l}^{(j)}(k) \right) P_{u_l t}^{\Phi} \, du_l$$

remembering that the transition probabilities of the phase process are denoted by P^{Φ}_{st}.
☺

Remark 2.13 The first moment formulae given in Pacheco and Prabhu [105], theorem 6.15 and corollary 6.16(a), can be obtained from this theorem as the special case of (N, J) being homogeneous, $\Sigma = I\!N_0^{\tau}$ and Φ being countable.

5. Laws of Large Numbers

For some special cases of Markov–additive jump processes on real vector spaces, the asymptotic behaviour can be described in terms of strong laws of large numbers. First, this will be done for processes with periodic generators (cf. chapter 1, section 5). The case of homogeneous processes is an obvious corollary of the periodic case. In this section, convergence on the vector space Σ shall be defined as convergence in every dimension of Σ.

Theorem 2.14 *Let (N, J) denote a periodic Markov–additive jump process with period T and be $(\Sigma, +)$ a real vector space with base $(b_j : j \in I)$. Let the phase process J have a periodic family of asymptotic distributions $(\pi_t : t \in [0, T[)$. Define the mean rate vector during a period length by*

$$\lambda := \left(\frac{1}{T}\int_0^T \int_\Phi d\pi_t(y) \int_{-\infty}^{\infty} c\, q_t(y, pr_j^{-1}(dc) \times \Phi)\, dt : j \in I\right)$$

Assume that $\|pr_j(E(N_t))\| < M(t) < \infty$ for all $j \in I$ and $t \in [0, T]$. Then

$$\frac{N_t}{t} \to \lambda \qquad as \quad t \to \infty$$

P-almost surely for all initial distributions.

Proof: Because of property (2.1), the evolution of the process (N, J) depends only on the initial distribution μ of J_0 on the phase space Φ. Since further for every initial value $N_0 \neq 0$ the quotient N_0/t vanishes for $t \to \infty$, we can assume $N_0 = 0$ almost surely without loss of generality. Let N_t^{μ} denote the marginal process N at time t under initial phase distribution μ. Further, let $\varphi_t = \mu P^{\Phi}(0, t)$ denote the distribution of the phase process at time t.

Choose any $\varepsilon > 0$. Then there is a number $k \in I\!N$ such that we can guarantee $\|\mu P^{\Phi}(0, nT) - \pi_0\| < \varepsilon$ for all $n \geq k$. The representation

$$\begin{aligned}
\frac{N_t}{t} &= \frac{1}{t}\left(N_{kT} + (N_{\lfloor t/T\rfloor T} - N_{kT}) + (N_t - N_{\lfloor t/T\rfloor T})\right) \\
&= \frac{1}{t}N_{kT} + \frac{1}{t}N^{\varphi_{kT}}_{(\lfloor t/T\rfloor - k)T} + \frac{1}{t}(N_t - N_{\lfloor t/T\rfloor T})
\end{aligned}$$

holds because of the periodicity of the process. Since the expectation kernel $E(N_s)$ is bounded for all $s \in I\!R_0$, the first and the last term will vanish for $t \to \infty$ almost surely. The second term equals

$$\begin{aligned}\frac{1}{t}N^{\varphi_{kT}}_{(\lfloor t/T \rfloor - k)T} &= \frac{1}{t}N^{\pi_0}_{(\lfloor t/T \rfloor - k)T} + \frac{1}{t}N^{\varphi_{kT}-\pi_0}_{(\lfloor t/T \rfloor - k)T} \\ &= \frac{1}{t}\sum_{m=k}^{\lfloor t/T \rfloor - 1} N^{\pi_0}_T + \frac{1}{t}N^{\varphi_{kT}-\pi_0}_{(\lfloor t/T \rfloor - k)T}\end{aligned}$$

since $\pi_0 = \pi_0 P^{\Phi}(0, T)$ is invariant. The first term of this sum tends to

$$\begin{aligned}\lim_{t\to\infty}\frac{1}{t}\sum_{m=k}^{\lfloor t/T \rfloor - 1} N^{\pi_0}_T &= \lim_{t\to\infty}\frac{(\lfloor t/T \rfloor - k)T}{t}\frac{1}{(\lfloor t/T \rfloor - k)T}\sum_{m=k+1}^{\lfloor t/T \rfloor} N^{\pi_0}_T \\ &= \lim_{t\to\infty}\frac{(\lfloor t/T \rfloor - k)T}{t}\cdot\lim_{n\to\infty}\frac{1}{(n-k)T}\sum_{m=k+1}^{n} N^{\pi_0}_T \\ &= \frac{1}{T}E(N^{\pi_0}_T) = \lambda\end{aligned}$$

almost surely, according to Kolmogorov's law of large numbers (cf. Bauer [15], p.86) and theorem 2.8. The assumption $\|\mu P^{\Phi}(0, kT) - \pi_0\| < \varepsilon$ implies

$$P\left(N^{\varphi_{kT}}_{(\lfloor t/T \rfloor - k)T} \neq N^{\pi_0}_{(\lfloor t/T \rfloor - k)T}\right) < \varepsilon$$

which means that

$$N^{\varphi_{kT}-\pi_0}_{(\lfloor t/T \rfloor - k)T} = N^{\varphi_{kT}}_{(\lfloor t/T \rfloor - k)T} - N^{\pi_0}_{(\lfloor t/T \rfloor - k)T} = 0$$

with probability $1 - \varepsilon$. Hence with probability $1 - \varepsilon$, the convergence $\frac{N_t}{t} \to \lambda$ holds. Since ε can be chosen arbitrarily small, there can be no set N with $P(N) > 0$ such that $\frac{N_t(\omega)}{t} \not\to \lambda$ for all $\omega \in N$.
☺

Remark 2.15 If there is a norm on Σ (e.g. the supremum norm, defined as $\|x\| := \sup_{j\in I} |pr_j(x)|$ for all $x \in \Sigma$) and Σ is complete with respect to this norm, i.e. if Σ is a Banach space, then the above theorem is valid for convergence in terms of the norm on Σ, too. This can be proven in exactly the same way by using the strong law of large numbers on Banach spaces (cf. Mourier [90] or Beck [18]) instead of Kolmogorov's law of large numbers, which is a statement for real-valued random variables.

Theorem 2.16 *Let (N, J) denote a homogeneous Markov–additive jump process and be $(\Sigma, +)$ a real vector space with base $(b_i : i \in I)$. Assume that the phase process J has an asymptotic distribution π. Define the mean rate vector by*

$$\lambda := \left(\int_\Phi d\pi(y) \int_{-\infty}^{\infty} c\, q(y, pr_j^{-1}(dc) \times \Phi) : j \in I \right)$$

If $\|pr_j(E(N_1))\| < M < \infty$ for all $j \in I$, then

$$\frac{N_t}{t} \to \lambda \qquad as \quad t \to \infty$$

P-almost surely for all initial distributions.

Proof: This follows immediately from theorem 2.14, since a homogeneous process with asymptotic distribution π is a periodic process with arbitrary period length $T > 0$ and periodic family $(\pi_t = \pi : t \in [0, T[)$ of asymptotic distributions. Furthermore,

$$\int_\Phi d\pi(y) \int_{-\infty}^{\infty} c\, q(y, pr_j^{-1}(dc) \times \Phi)$$
$$= \frac{1}{T} \int_0^T \int_\Phi d\pi_t(y) \int_{-\infty}^{\infty} c\, q_t(y, pr_j^{-1}(dc) \times \Phi)\, dt$$

holds for all $i \in I$, since $q_t(y, pr_j^{-1}(dc) \times \Phi) = q(y, pr_j^{-1}(dc) \times \Phi)$ is constant in t.
☺

Remark 2.17 The statement of this theorem reduces to the result in Pacheco and Prabhu [105], theorem 6.17, if one specifies $\Sigma = I\!N_0^r$ and Φ being countable.

II

CLASSICAL QUEUES

Chapter 3

EXAMPLES OF MARKOVIAN ARRIVAL PROCESSES

The present chapter provides two important examples for Markovian arrival processes. In contrast to the top–down approach in chapter 2, where arrival processes have been introduced from a very general point of view, the following presentation shall provide some intuition into the common features of different specifications of MAPs. Therefore, we begin with the simple example (which is the well–known homogeneous BMAP) and proceed with generalizations in two respects: First, the processes are developed to include inhomogeneity in time. Second, the arrival space is generalized to a continuous one.

After this introduction of special classes of MAPs, the next chapters contain the analysis of queues with (possibly inhomogeneous) BMAP inputs. Although there will be no examination of queues with FMAP input, the analogies that will be shown between BMAPs and FMAPs should make clear how to translate the methods to the analysis of FMAP queues.

1. Batch Markovian Arrival Processes

As a first application of the results derived in the first part of the book, the concept of a Batch Markovian Arrival Process shall be described and generalized within the framework of MAJPs. Markovian Arrival Processes (MAPs) and Batch Markovian Arrival Processes (BMAPs) have been introduced by Neuts [95] and Lucantoni et al. [82, 80] in order to provide input streams for queueing systems which are Markovian (and hence analytically more tractable) on the one hand but very versatile (even dense in the class of point processes, see Asmussen and Koole [6]) on the other hand. This concept has proved very successful in queueing theory for more than twenty years now. For a bibliography demonstrating this, see Lucantoni [81].

1.1 The homogeneous case

The classical homogeneous BMAP as introduced in Lucantoni [80] is a homogeneous MAJP with additive space $\Sigma = \mathbb{N}_0$ and finite phase space $\Phi = \{1, \ldots, m\}$. As a homogeneous process, its generator function is constant. Because of the low complexity of the state space, we can represent the generator in the usual block matrix form as

$$Q = \begin{pmatrix} D_0 & D_1 & D_2 & D_3 & \cdots \\ & D_0 & D_1 & D_2 & \cdots \\ & & D_0 & D_1 & \ddots \\ & & & D_0 & \ddots \\ & & & & \ddots \end{pmatrix}$$

with $m \times m$ matrices $\Delta = (D_n : n \in \mathbb{N})$. Because of its self–similarity (see remark 2.2), the sequence of matrices Δ contains all information for Q and thus may be called the **characteristic sequence** of the BMAP.

Since the state space $S = \mathbb{N}_0 \times \{1, \ldots, m\}$ is countable, the transition kernel from any time s to some later time $t > s$ is a matrix of infinite dimension. Denote this by P_{st}. According to theorem 1.13, it is given by

$$P_{st} = e^{Q \cdot (t-s)} = \sum_{n=0}^{\infty} \frac{(t-s)^n}{n!} Q^n$$

for all $s < t$, with Q^n denoting the nth power of the matrix Q.

Because of the upper triagonal form of the block matrix Q, the computation of the transition matrix can be performed by using the convolution calculus introduced in Baum [16]. This shall shortly be presented as follows: First we can write $P(t-s)$ instead of P_{st} for all $s < t$, since the process is homogeneous. Thus $(P(t) : t > 0)$ contains all informations for the distribution of the homogeneous BMAP. Now the block matrices $P(t)$ have the same upper triagonal form as Q, since a BMAP is an arrival process. Because of the homogeneity in the first dimension, we can even write

$$P(t) = \begin{pmatrix} P_0(t) & P_1(t) & P_2(t) & P_3(t) & \cdots \\ & P_0(t) & P_1(t) & P_2(t) & \cdots \\ & & P_0(t) & P_1(t) & \ddots \\ & & & P_0(t) & \ddots \\ & & & & \ddots \end{pmatrix}$$

for all $t > 0$, with $m \times m$ matrices $P_n(t)$ having entries

$$P_{n;ij}(t) := (P_n(t))_{ij} := P_{0t}((0,i),(n,j))$$

for all $n \in \mathbb{N}$ and $1 \leq i, j \leq m$.

For two sequences $A = (A_n : n \in \mathbb{N}_0)$ and $B = (B_n : n \in \mathbb{N}_0)$ of $m \times m$ matrices, define the convolution sequence $C = (C_n : n \in \mathbb{N}_0) = A * B$ by $C_n := \sum_{i=0}^{n} A_i B_{n-i}$ for all $n \in \mathbb{N}_0$. Based on this define the convolutional power of a sequence of matrices by $A^{*0} := (\delta_{0,n} \cdot I : n \in \mathbb{N}_0)$, with I denoting the identity matrix, and recursively $A^{*n+1} := A^{*n} * A$ for all $n \in \mathbb{N}_0$. Then the structure of the matrix Q leads to a simple representation

$$P_k(t) = e^{*\Delta \cdot t} := \sum_{n=0}^{\infty} \frac{t^n}{n!} \Delta_k^{*n} \tag{3.1}$$

for all $k \in \mathbb{N}_0$ and $t > 0$, with Δ_k^{*n} denoting the kth matrix in the sequence Δ^{*n}. All calculations in this expression require only finite–dimensional matrices, which makes it suitable for computation.

Concluding this section, we shall specify the results for expectations and asymptotic behaviour. Define the stochastic $m \times m$ matrix $D := \sum_{n=0}^{\infty} D_n$. According to equation (2.6), this is the generator of the phase process. Let π be a stationary probability vector for D, such that $\pi D = 0$ holds. Now theorems 2.12 and 1.13 yield for the expectation kernel (which is an $m \times m$ matrix here) the expression

$$E(N_t - N_s) = \int_s^t e^{D \cdot (u-s)} \sum_{n=1}^{\infty} n D_n \, e^{D \cdot (t-u)} \, du$$

for all $s < t$. If the process starts without prior arrivals and in phase equilibrium π, this becomes

$$E_\pi(N_t) = \pi \sum_{n=1}^{\infty} n D_n 1_m \cdot t$$

for all $t > 0$, with 1_m denoting the m–dimensional column vector with all entries being one. This simplification is due to the properties

$$\pi e^{D \cdot (u-s)} = \pi \qquad \text{and} \qquad e^{D \cdot (t-u)} 1_m = 1_m \tag{3.2}$$

which hold for all $s < u < t$. Finally, theorem 2.16 yields the asymptotics

$$\frac{N_t}{t} \to \pi \sum_{n=1}^{\infty} n D_n 1_m \qquad \text{as} \quad t \to \infty$$

P–almost surely for all initial phase distributions. The term $\pi \sum_{n=1}^{\infty} n D_n 1_m$ is called stationary arrival rate.

1.2 The general case

As an obvious extension of the classical BMAP concept, a general (i.e. possibly inhomogeneous) BMAP shall be defined as a (possibly inhomogeneous) MAJP with additive space $\Sigma = I\!N_0$ and finite phase space $\Phi = \{1, \ldots, m\}$. Because of the state space $E = I\!N_0 \times \{1, \ldots, m\}$ we have the same structure of the generators

$$Q_t = \begin{pmatrix} D_0(t) & D_1(t) & D_2(t) & D_3(t) & \cdots \\ & D_0(t) & D_1(t) & D_2(t) & \cdots \\ & & D_0(t) & D_1(t) & \ddots \\ & & & D_0(t) & \ddots \\ & & & & \ddots \end{pmatrix}$$

as in the homogeneous case, only that the characteristic sequences $\Delta(t) = (D_n(t) : n \in I\!N_0)$ depend on time. General BMAPs keep the most important structural properties of homogeneous BMAPs, such that the computations necessary for analyzing them and the respective queues are still tractable.

As usual, an entry $D_n(t)(i, j)$ can be interpreted as the infinitesimal transition rate of observing n arrivals during the infinitesimal time interval dt while changing from phase i to phase j. Likewise, all the other interpretations for BMAPs can easily be adapted to the time–inhomogeneous case. Denote in the following a BMAP by $(N, J) = ((N_t, J_t) : t \in I\!R_0^+)$, meaning that N_t denotes the number of arrivals until time t and J_t denotes the arrival phase at any time t.

In order to determine the transition probabilities of a general BMAP (N, J), a generalization of the convolution calculus developed in Baum [16] proves useful. Remember that $P_{st}(i, (k, j))$ is the probability of observing $k \in I\!N_0$ arrivals and being in phase j at time $t > s$ under the condition of having observed 0 arrivals and being in phase i at time s. Define $P_k(s, t) = (P_{st}(i, (k, j)))_{i,j \in \Phi}$ as the $m \times m$ matrix with entries $P_{st}(i, (k, j))$ as well as the sequence of matrices $P(s, t) = (P_k(s, t) : k \in I\!N_0)$. Then definition (2.3) is specified to

$$P_k^{(0)}(s, t) := \delta_{k,0} \cdot I$$

for all $s < t \in I\!R_0^+$, with I denoting the identity matrix on Φ. Because of the upper triagonal form of the block matrices Q_u, the recursion (2.4) translates to

$$P_k^{(n)}(s, t) := \sum_{l=0}^{k} \int_s^t P_l^{(n-1)}(s, t) D_{k-l}(u)\, du$$

for all $k, n \in \mathbb{N}_0$. With these, the transition matrices are given by

$$P_k(s,t) = \sum_{n=0}^{\infty} P_k^{(n)}(s,t)$$

for all $s < t \in \mathbb{R}_0^+$ and $k \in \mathbb{N}_0$. Analogous to formula (2.5), the sequences $P^{(n)}(s,t) = (P_k^{(n)}(s,t) : k \in \mathbb{N}_0)$ can be expressed in closed form as

$$P^{(n)}(s,t) = \underbrace{\int_s^t \int_s^{u_n} \dots \int_s^{u_2}}_{n \ integrals} \Delta(u_1) * \dots * \Delta(u_n) \, du_1 \dots du_n \tag{3.3}$$

for every $n \in \mathbb{N}_0$ and $s < t \in \mathbb{R}_0^+$, using the convolution $*$ as defined in the previous section.

Analogously to remark 1.9, the sequence P_{st} of transition probability matrices can be computed iteratively by starting with $I_0(s,t) := P_{st}^{(0)}$ and iterating by

$$I_{n+1}(s,t) := \int_s^t I_n(s,u) * \Delta(u) \, du + I_0(s,t)$$

for $n \in \mathbb{N}_0$. Then $P_{st} = \lim_{n\to\infty} I_n(s,t)$ for all $s < t \in \mathbb{R}_0^+$.

As in the homogeneous case, the $m \times m$ matrix $D(t) := \sum_{n=0}^{\infty} D_n(t)$ is a generator for all $t \in \mathbb{R}_0^+$. Again, the MJP with generator function $t \to D(t)$ shall be called the **phase process** of the BMAP. Denote the transition matrix of the phase process from time s to time $t \geq s$ by $P_{s,t}^{\Phi}$. Then theorem 2.12 yields the expression

$$E(N_t - N_s) = \int_s^t P_{su}^{\Phi} \sum_{k=1}^{\infty} k \cdot D_k(u) \, P_{ut}^{\Phi} \, du \tag{3.4}$$

for the expectation matrix of the additive process N over the time interval $]s,t]$. If the phase process has a stationary distribution π such that $\pi = \pi P_{0t}^{\Phi}$ for all $t \in \mathbb{R}^+$, then starting the phase process in this distribution without prior arrivals leads to expectation matrices

$$E_\pi(N_t) = \int_0^t \pi \sum_{k=1}^{\infty} k D_k(u) 1_m \, du$$

for the number of arrivals until time $t \in \mathbb{R}^+$.

A **periodic BMAP** with period T shall be defined as a general BMAP with the property $\Delta(s+T) = \Delta(s)$ for all $s \in [0,T[$. If the phase process of a periodic BMAP has a periodic family $(\pi_t : t \in [0,T[)$ of asymptotic distributions, then theorem 2.14 yields

$$\frac{N_t}{t} \to \frac{1}{T} \int_0^T \pi_t \sum_{n=1}^{\infty} n D_n(t) 1_m \, dt \qquad \text{as} \quad t \to \infty$$

P-almost surely for all initial distributions.

2. Fluid Markovian Arrival Processes

There is a strong connection between queueing theory and the theory of risk and storage processes (see Prabhu [107] and Asmussen [5]). This link gives rise to queueing models with continuous state spaces, the so–called fluid queues. However, for such a concept a more general class of arrival processes is needed. This shall be presented in the present section.

If we generalize the concept of a characteristic sequence slightly, we will obtain a class of Markovian arrival processes which is suitable for fluid queueing models. Again, assume that the phase space is finite, i.e. let $\Phi = \{1, \dots, m\}$. For arrivals into fluid queues, we need an additive space $\Sigma = I\!R_0^+$. As in the preceding section, we will first introduce the homogeneous case as an illustration, and after that describe the general, possibly inhomogeneous case.

2.1 The homogeneous case

For BMAPs, which have an additive space $I\!N_0$, it was possible to arrange the matrices containing the arrival rates in a sequence, called the characteristic sequence. If the additive space is $I\!R_0^+$ as in the present case, an analogue would be a characteristic measure which provides an arrival rate matrix for every Borel–measurable subset of $I\!R_0^+$.

Thus we come to the following definitions: A homogeneous MAJP with phase space $\Phi = \{1, \dots, m\}$ and additive space $\Sigma = I\!R_0^+$ shall be called homogeneous Fluid Markovian Arrival Process (or shortly: homogeneous FMAP). As such its distribution is completely determined by the infinitesimal transition rates $q(i; A \times \{j\})$ with $A \in \mathcal{B}_0^+$ and $i, j \in \Phi$, see the definitions after (2.2). Furthermore, since $\mathcal{B}_0^+$ is generated by sets of the form $[0, x]$ with $x \in I\!R_0^+$, it suffices to know the rates $q(i; [0, x] \times \{j\})$. Thus the characteristic measure Δ of a homogeneous FMAP shall be defined by

$$\Delta(x) := (q(i; [0, x] \times \{j\}))_{i,j \in \Phi}$$

for all $x \in I\!R_0^+$. Thus $\Delta(x)$ is an $m \times m$ matrix, with a completely analogous meaning as the D_n for BMAPs.

Remark 3.1 The term characteristic measure has been chosen because for any given $i, j \in \Phi$, the (i, j)th entry $\Delta(.)(i, j)$ of the matrix–valued function $\Delta(.)$ is the distribution function of a sub–stochastic measure on the additive space

$(\mathbb{R}_0^+, \mathcal{B}_0^+)$, according to the representation in theorem 2.5. For BMAPs, the characteristic sequence $(D_n : n \in \mathbb{N}_0)$ constitutes a measure on the discrete space $\mathbb{N}_0$, too.

The homogeneity in the first dimension of the state space enables us to find concise expressions for the transition probabilities in terms of convolutional matrix exponential forms again. To this aim define $\Delta^{*0}(x) := I$ for all $x \in \mathbb{R}_0^+$, with I denoting the identity matrix, and recursively

$$\Delta^{*n+1}(x) := \int_0^x \Delta^{*n}(x - z)\, d\Delta(z)$$

for all $x \in \mathbb{R}_0^+$ and $n \in \mathbb{N}_0$. Here, $\Delta(.)$ denotes the matrix–valued function for which the (i,j)th entry is the distribution function $\Delta(.)(i,j)$, see the remark above.

Since we observe a homogeneous process, it suffices to find expressions for the transition kernels $P_t := P_{0,t}$ in order to determine its distribution. Then according to theorem 1.13 and expression (2.5), we obtain

$$P_t(x) = e^{*\Delta \cdot t}(x) = \sum_{n=0}^{\infty} \frac{t^n}{n!} \Delta^{*n}(x)$$

for all $x \in \mathbb{R}_0^+$, which is an obvious analogue to formula (3.1).

As in the previous section, we shall conclude by specifying the results for expectations and asymptotic behaviour. The generator of the phase process is given here by $D := \lim_{x\to\infty} \Delta(x)$. Let π be a stationary probability vector for D, such that $\pi D = 0$ holds. Theorems 2.8 and 1.13 then yield for the expectation kernel (which again is an $m \times m$ matrix here) the expression

$$E(N_t - N_s) = \int_s^t e^{D\cdot(u-s)}\, E(\Delta)\, e^{D\cdot(t-u)}\, du$$

for all $s < t$, with $E(\Delta)$ denoting the $m \times m$ matrix with (i,j)th entry $\int_0^\infty z \; d\Delta(z)(i,j)$. If the process starts without prior arrivals and in phase equilibrium π, this becomes

$$E_\pi(N_t) = \pi\, E(\Delta)\, 1_m \cdot t$$

for all $t > 0$, with 1_m denoting the m–dimensional column vector with all entries being one. As in the BMAP example, this simplification is due to properties (3.2). Theorem 2.16 yields the asymptotics

$$\frac{N_t}{t} \to \pi E(\Delta) 1_m \qquad \text{as} \quad t \to \infty$$

P–almost surely for all initial phase distributions.

2.2 The general case

Analogously to the generalization from homogeneous BMAPs to general BMAPs, a general FMAP shall be defined as a possibly inhomogeneous MAJP with phase space $\Phi = \{1, \ldots, m\}$ and additive space $\Sigma = I\!R_0^+$. For every $t \in I\!R_0^+$, define the characteristic measure at time t by

$$\Delta_t(x) := (q_t(i, [0, x] \times \{j\}))_{i,j \in \Phi}$$

for all $x \in I\!R_0^+$. Again, $\Delta_t(x)$ is an $m \times m$ matrix and plays an analogous role as the matrix $D_n(t)$ for general BMAPs.

The possible inhomogeneity in time does not destroy the homogeneity in the additive space. Hence we can derive closed form expressions for the transition probabilities via matrix convolutions. For this, define a convolution of characteristic measures by

$$(\Delta_u * \Delta_v)(x) := \int_0^x \Delta_v(x - z) d\Delta_u(z)$$

for all $u, v \in I\!R_0^+$ and $x \in I\!R_0^+$, again with $\Delta_u(.)$ interpreted as a matrix of distribution functions. Then $(\Delta_u * \Delta_v)(x)$ is an $m \times m$ matrix for every $x \in I\!R_0^+$. By induction, it is clear that

$$(\Delta_{u_1} * \ldots * \Delta_{u_{n+1}})(x) := ((\Delta_{u_1} * \ldots * \Delta_{u_n}) * \Delta_{u_{n+1}})(x)$$

is again an $m \times m$ matrix for every $x \in I\!R_0^+$ and $n \in I\!N_0$.

With these definitions, expression (2.3) is specified to $P_{s,t}^{(0)}(x) = I$ for all $x \in I\!R_0^+$, with I denoting the identity matrix. Recursion (2.4) is now given by

$$P_{s,t}^{(n+1)}(x) = \int_s^t \int_0^x P_{s,u}^{(n)}(dz)\Delta_u(x - z)\, du$$

for all $s < t$, $x \in I\!R_0^+$ and $n \in I\!N_0$. For the transition probabilities from time s to time t, formula (2.5) is specified to

$$P_{s,t}^{(n)}(x) = \underbrace{\int_s^t \int_s^{u_n} \cdots \int_s^{u_2}}_{n \text{ integrals}} (\Delta_{u_1} * \ldots * \Delta_{u_n})(x)\, du_1 \ldots du_n$$

for all $x \in I\!R_0^+$.

The phase process of an FMAP is determined by its generator function $t \to D_t$ with

$$D_t = \lim_{x \to \infty} \Delta_t(x)$$

for all $t \geq 0$. Denote the transition matrix of the phase process from time s to time $t \geq s$ by $P^{\Phi}_{s,t}$. Then theorem 2.8 yields the expression

$$E(N_t - N_s) = \int_s^t P^{\Phi}_{su}\, E(\Delta_u)\, P^{\Phi}_{ut}\, du$$

for the expectation matrix of the additive process N over the time interval $]s,t]$. Analogous to the homogeneous case, $E(\Delta_u)$ denotes the matrix with (i,j)th entry $\int_0^\infty z\, d\Delta_u(z)(i,j)$.

If the phase process has a stationary distribution π such that $\pi = \pi P^{\Phi}_{0t}$ for all $t \in I\!R^+$, then starting the phase process in this distribution without prior arrivals leads to expectation matrices

$$E_\pi(N_t) = \int_0^t \pi\, E(\Delta_u)\, 1_m\, du$$

for the number of arrivals until time $t \in I\!R^+$.

A periodic FMAP with period T shall be defined as a general FMAP with the property $\Delta_{s+T} = \Delta_s$ for all $s \in [0,T[$. If the phase process of a periodic FMAP has a periodic family $(\pi_t : t \in [0,T[)$ of asymptotic distributions, then theorem 2.14 yields

$$\frac{N_t}{t} \to \frac{1}{T}\int_0^T \pi_t\, E(\Delta_t)\, 1_m\, dt \qquad \text{as} \quad t \to \infty$$

P-almost surely for all initial distributions.

Chapter 4

THE PERIODIC $BMAP/PH/C$ QUEUE

In queueing theory, most models are based on time-homogeneous arrival processes and service time distributions. One of the most important features to be exploited is the Markov property which often appears after the construction of embedded Markov chains. The search for Markovian but versatile arrival processes has led to the concept of batch Markovian arrival processes (BMAPs) which allow for a phase process controlling the arrival rates. This arrival process is often used for modelling communication networks.

A typical property of communication traffic is the dependence of its arrival rates on time. This aspect incites the use of time–inhomogeneous processes and queues for modelling communication networks. Typically, a periodic dependence of the arrival rates and/or the service time distribution can be assumed with period lengths of a day or a week.

While queues with periodic input naturally reflect the time–dependence of the traffic that arrives in communication networks, the analysis of queues with inhomogeneous arrival rates is far less developed than the one for homogeneous queues. Some of the existing results in the literature are given in Asmussen and Thorisson [8], Bambos and Walrand [13], Falin [57], Harrison and Lemoine [65], Hasofer [66], Heyman and Whitt [68], Lemoine [79, 78], Massey [83], Rolski [112, 113], and Willie [128]. Although many types of stability conditions could be established, the first explicit formulae for asymptotic behaviour could be derived only recently and for Markovian queues, see Breuer [29, 32].

The present chapter is organized as follows. In the remainder of this introduction, the periodic $BMAP/PH/c$ queue as well as the basic notations are defined. The first section contains the transient distributions at arbitrary times. In section 2, a stability condition is given and the asymptotic distributions at any "day time" of one period length are derived explicitly in the case of stability.

Let $Q = (Q_t : t \in I\!R_0^+)$ denote a periodic $BMAP/PH/c$ queue with a periodic BMAP arrival process and a phase–type service time distribution which is identical and independent for all c servers. Define the arrival process by its transition rate matrices $(D_n(t) : n \in I\!N_0, t \in I\!R_0^+)$ having dimension m and period T.

Every server shall be equal, and the service time distribution be phase–type with representation (α, S) and dimension r (for a description of phase–type distributions, see Neuts [98]). The absorbing state shall be denoted by 0 and the transient states by $\{1, \ldots, r\}$. For any $d \in I\!N$, let 1_d denote the d–dimensional column vector with all entries being 1 and denote the ith canonical column base vector of $I\!R^d$ by e_i if the dimension d is clear from the context. Denote the transpose of a matrix A or a vector v by A^T or v^T, respectively. Finally, define the exit vector $\eta := -S1_r$, the vectors $\alpha' := (0, \alpha)$ and $\eta' := (0, \eta^T)^T$ as well as the matrices

$$S' := \begin{pmatrix} 0 & \ldots 0 \\ \vdots & S \\ 0 & \end{pmatrix}, \qquad B_0 := \begin{pmatrix} \alpha' \\ 0 \ldots 0 \\ \vdots \\ 0 \ldots 0 \end{pmatrix} \qquad \text{and} \qquad B_1 := \begin{pmatrix} 0 & \alpha \\ \vdots & I \\ 0 & \end{pmatrix}$$

The multiplication of two $I\!N_0 \times I\!N_0$ block matrices A and B with blocks of dimension d is defined by

$$(AB)_{kn}(i, j) := \sum_{l=0}^{\infty} \sum_{h=1}^{d} A_{kl}(i, h) B_{ln}(h, j) \tag{4.1}$$

for every $k, n \in I\!N_0$ and $i, j \in \{1, \ldots, d\}$.

1. Transition Probabilities

As a Markovian queue, the system can be analyzed like a (periodic) Markov jump process. The process Q has state space $I\!N_0 \times \{1, \ldots, m\} \times \{0, \ldots, r\}^c$ and hence a finite phase space of dimension $1 + c$ with $d := m \cdot (r + 1)^c$ possible phases. The first dimension indicates the arrival phase and the last c dimensions shall describe the phases of the respective servers. If a server is in phase 0 at time t, it means that this server is idle at that time.

Let 0 and I denote the zero and identity matrix, respectively. Denote the Kronecker product of two quadratic matrices A and B by $A \otimes B$ (cf. Bellman [19]). The neutral element with respect to the Kronecker product is the scalar $1 \in I\!R$, interpreted as a one–dimensional matrix. Define the iteration of Kronecker products by $A^{\otimes 0} := 1$ and $A^{\otimes n+1} := A^{\otimes n} \otimes A$ for all quadratic matrices A.

Now the infinitesimal generator $G(t)$ of the queue process Q can be written as an $\mathbb{N}_0 \times \mathbb{N}_0$ block matrix with entries being the $d \times d$ matrices

$$G_{kn}(t) = \begin{cases} 0, & k > n+1 \\ \sum_{i=1}^{c} I^{\otimes i} \otimes \eta' e_0^T \otimes I^{\otimes c-i}, & k = n+1 \leq c \\ \sum_{i=1}^{c} I^{\otimes i} \otimes \eta' \alpha' \otimes I^{\otimes c-i}, & k = n+1 > c \\ D_0(t) \otimes I^{\otimes c} + \sum_{i=1}^{c} I^{\otimes i} \otimes S' \otimes I^{\otimes c-i}, & k = n \\ D_{n-k}(t) \otimes I^{\otimes c}, & c \leq k < n \\ D_{n-k}(t) \otimes B_1^{\otimes c}, & k < c \leq n \end{cases}$$

as well as

$$G_{kn}(t) = \binom{c-k}{c-n}^{-1} \sum_{1 \leq i_1 < \ldots < i_{n-k} \leq c} D_{n-k}(t) \otimes M_1 \otimes \ldots \otimes M_c$$

for $k < n < c$, with $M_i = B_0$ for $i \in \{i_1, \ldots, i_{n-k}\}$ and $M_i = I$ else. The interpretation of the last equation is that for several possibilities of filling idle servers, every possibility shall be equally probable. This convention allows us to write the above generator in block matrix form without needing to define the elements of the blocks separately, which notationally would be much more inconvenient. Denote the (i,j)th entry of the matrix $G_{kn}(t)$ by $G_{kn}(t)(i,j)$ for $i,j \in \{1, \ldots, d\}$.

Remark 4.1 Instead of the periodic $BMAP/PH/c$ queue as described above one can analyze the periodic $BMAP/M_t/c$ queue with periodically varying service rates by the same method. In this case, every server is equal, and the service time distribution function B_s for a user arriving at time $s \in \mathbb{R}_0^+$ is defined by

$$B_s(t-s) := 1 - e^{-\int_s^t \mu_u du}$$

for all $t > s$. This means that the service process without idle periods would be an inhomogeneous Poisson process with rates $(\mu_t : t \in \mathbb{R}_0^+)$. Periodicity of the service rates means $\mu_{s+T} = \mu_s$ for all $s \in [0, T[$.

Here, we would have an infinitesimal generator $G(t)$ of the queue process which can be written as an $\mathbb{N}_0 \times \mathbb{N}_0$ block matrix with entries being the $m \times m$ matrices

$$G_{kn}(t) = \begin{cases} 0, & k > n+1 \\ k\mu_t \cdot I, & k = n+1 \leq c \\ c\mu_t \cdot I, & k = n+1 > c \\ D_0(t) - k\mu_t \cdot I, & k = n \leq c \\ D_0(t) - c\mu_t \cdot I, & k = n > c \\ D_{n-k}(t), & k < n \end{cases}$$

for $k, n \in \mathbb{N}_0$. All the statements in the following apply to this queue with periodic service rates, too.

Define $P_{st}(k, i; n, j)$ as the probability of having $n \in \mathbb{N}_0$ users in the queue and being in phase j at time $t > s$ under the condition of having $k \in \mathbb{N}_0$ users in the queue and being in phase i at time s. Further define $P_{st}(k, n)$ as the $d \times d$ matrix with entries $P_{st}(k, i; n, j)$ and $P_{st} = (P_{st}(k, n))_{k,n \in \mathbb{N}_0}$ as the $\mathbb{N}_0 \times \mathbb{N}_0$ block matrix with entries $P_{st}(k, n)$.

Obviously, the queueing process is a Markov jump process with a periodic generator function as given above. Hence the transition probability matrices P_{st} are given by

$$P_{st} = \sum_{k=0}^{\infty} \underbrace{\int_s^t \int_s^{u_k} \dots \int_s^{u_2}}_{k\ integrals} G(u_1) \dots G(u_k) du_1 \dots du_k$$

for all $s < t$, according to theorem 1.12 and equation (1.6). The periodicity allows the same simplifications as for periodic MJPs in chapter 1. Furthermore, the iteration given in equation (1.7) yields a way of computing the remaining terms.

2. Stability and Asymptotic Distributions

Before asymptotic distributions can be given, the asymptotic analysis of the $BMAP/PH/c$ queue first requires the derivation of an ergodicity condition. In Fayolle et al. [58], a version of Foster's criterion for state spaces $\mathbb{N}_0 \times \{1, \dots, d\}$, hence for Markovian BMAP queues with phase–type service, can be found. This shall be used in this section in order to prove ergodicity criteria for the periodic $BMAP/PH/c$ queue. After that, the concept of an asymptotic distribution (which is valid for homogeneous queues) is adapted to periodic queues by the definition of a periodic family of asymptotic distributions. This is given explicitly for the case of ergodicity at the end of this section.

Define $Y = (Y_n : n \in \mathbb{N}_0)$ as the homogeneous Markov chain with transition probability matrix $P_{0,T}$ and let $Y^\mu = (Y_n^\mu : n \in \mathbb{N})$ denote the version of Y with initial distribution μ. Assume in the following that the arrival phase process as well as the service phase processes have stationary distribution π_A and π_B, respectively. This means, the equations $\pi_A \sum_{n=0}^{\infty} D_n = 0$ and $\pi_B(S + \eta\alpha) = 0$ are satisfied. For the latter as well as for all following statements on PH–renewal processes, see Neuts [98], p.231ff.

Theorem 4.2 *The Markov chain Y is ergodic if and only if the stability condition*

$$\frac{1}{T}\int_0^T \pi_A \sum_{n=1}^{\infty} nD_n(t)1_m \, dt < c \cdot \pi_B \eta \tag{4.2}$$

holds.

Proof: Let $A = (A_t : t \in I\!R_0^+)$ and $B = (B_t : t \in I\!R_0^+)$ denote the BMAP arrival process into the queue and the c-fold superposition of the PH–renewal process with representation (α, S), respectively. That means, A_t is the random variable of all arrivals into the queue until time t. Define $Z := A - B$ as the difference of these independent processes. Then Z is a periodic Markov jump process. For any two–dimensional process X, denote the marginal process in the first dimension by X^1. The mean expectation of Z^1 over one period length in phase equilibrium equals

$$E(Z_T^1) = E(A_T^1) - E(B_T^1) = \int_0^T \pi_A \sum_{n=1}^{\infty} nD_n(t)1_m \, dt - c \cdot \pi_B \eta \cdot T \tag{4.3}$$

In order to show the necessity of condition (4.2), assume that $E(Z_T^1) \geq 0$. Since the state space of Z is $Z\!\!Z \times \{1, \ldots, d\}$, but the chain Y has a barrier at the zero level, we have

$$E(Y_1^1) > E(Z_T^1) \geq 0$$

for initial distributions with support $\{0\} \times \{1, \ldots, d\}$. Starting in phase equlibrium, the asymptotic expectation

$$\lim_{n\to\infty} E(Y_n^1) = \lim_{n\to\infty} \sum_{k=1}^{n} E(Y_1^1) = \infty$$

diverges to infinity. Hence, there is no asymptotic distribution for Y.

Now we show sufficiency. Denote the transition probability matrix of the homogeneous Markov chain $(Z_{nT} : n \in I\!N_0)$ by p^Z. Since Z is homogeneous in the first component, we can define

$$p_k^Z(i,j) := p_{(0,i),(k,j)}^Z = P(Z_T = (k,j) | Z_0 = (0,i))$$

for all $k \in Z\!\!Z$ and $i, j \in \{1, \ldots, d\}$. Further define $\pi := \pi_A \otimes \pi_B^{\otimes c}$, using the obvious adaptation of the Kronecker product to vectors. The above observation (4.3) yields

$$\sum_{i=1}^{d} \pi_i \sum_{k \in Z\!\!Z} \sum_{j=1}^{d} k \cdot p_k^Z(i,j) < 0$$

According to Fayolle et al. [58], p.35, there is an $\varepsilon > 0$ and a positive function f such that

$$\sum_{n=0}^{\infty}\sum_{j=1}^{d} p^Z_{(k,i),(n,j)} \cdot f(n,j) - f(k,i) < -\varepsilon$$

for almost all states $(k,i) \in I\!N_0 \times \{1,\ldots,d\}$. Furthermore, there are numbers $a_1,\ldots,a_d$ such that $f(k,i) = k + a_i$ for almost all states (k,i).

Define the event

$$R(n) := \{\exists\, t \in [nT,(n+1)T[: Q_t = 0\}$$

for all $n \in I\!N_0$. Then the transition probabilities of the homogeneous chain Y can be decomposed in

$$\begin{aligned}
p^Y_{(k,i),(l,j)} &:= P(Y_{n+1} = (l,j)|Y_n = (k,i)) \\
&= P(Y_{n+1} = (l,j)|Y_n = (k,i), R(n)) \cdot P(R(n)|Y_n = (k,i)) \\
&\quad + P(Y_{n+1} = (l,j)|Y_n = (k,i), R(n)^c) \cdot P(R(n)^c|Y_n = (k,i)) \\
&= P(Y_{n+1} = (l,j)|Y_n = (k,i), R(n)) \cdot P(R(n)|Y_n = (k,i)) \\
&\quad + p^Z_{(k,i)(l,j)} \cdot P(R(n)^c|Y_n = (k,i))
\end{aligned}$$

for all $n \in I\!N_0$. Since

$$\lim_{k\to\infty} P(R(n)|Y_n = (k,i)) = 0$$

there is a $k_0 \in I\!N$ such that

$$P(R(n)|Y_n = (k,i)) < \frac{\varepsilon}{2 \cdot M'}$$

for all $k > k_0$, with $M' := M + \sum_{i=1}^{d} |a_i|$ and some $M \in I\!N$, for which condition (4.2) implies

$$\max_{i\in\{1,\ldots,m\}} \int_0^T e_i \sum_{n=1}^{\infty} nD_n(t)1_m \, dt < M < \infty$$

Thus the estimation

$$\begin{aligned}
&\sum_{l=0}^{\infty}\sum_{j=1}^{d} P(Y_{n+1} = (l,j)|Y_n = (k,i), R(n)) \cdot f(l,j) \\
&\qquad \le \sum_{l=0}^{\infty}\sum_{j=1}^{d} P(Y_{n+1} = (l,j)|Y_n = (k,i), R(n)) \cdot \left(l + \sum_{j=1}^{d} |a_j|\right) \\
&\qquad < M + \sum_{i=1}^{d} |a_i|
\end{aligned}$$

holds. Using the positive function f, we have

$$\sum_{n=0}^{\infty}\sum_{j=1}^{d} p^{Y}_{(k,i),(n,j)} \cdot f(n,j) - f(k,i)$$
$$< \sum_{n=0}^{\infty}\sum_{j=1}^{d} p^{Z}_{(k,i),(n,j)} \cdot f(n,j) - f(k,i) + \frac{\varepsilon}{2 \cdot M'} \cdot \left(M + \sum_{i=1}^{d} |a_i| \right)$$
$$< -\frac{\varepsilon}{2}$$

for almost all states $(k,i) \in I\!N_0 \times \{1,\ldots,m\}$. Now Foster's criterion as stated in Fayolle et al. [58], p.29, assures that Y is ergodic.
☺

Now the concept of an asymptotic distribution shall be adapted to periodic queues. After that, the main theorem of this chapter gives an explicit formula for the asymptotic distribution of a stable periodic $BMAP/PH/c$ queue.

A family $(q_s : s \in [0,T[)$ of probability distributions shall be called a periodic family of asymptotic distributions for Q if

$$q_s = \lim_{n\to\infty} Q_{nT+s}$$

does exist for every $s \in [0,T[$. Here, the limit shall be defined in terms of weak convergence.

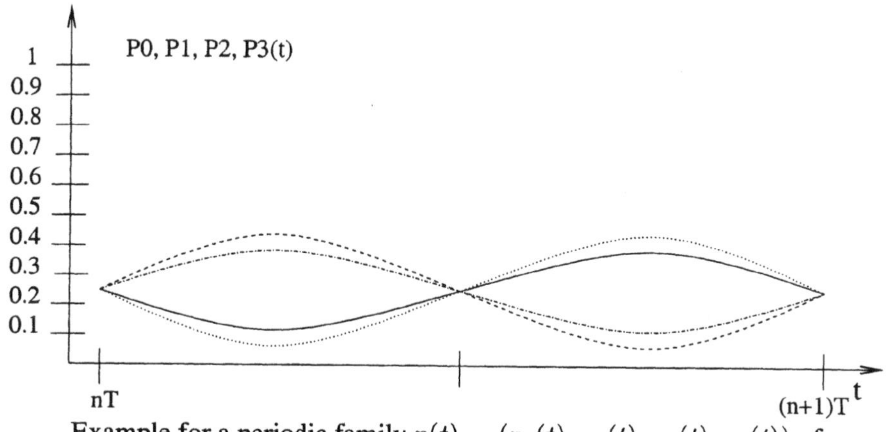

Example for a periodic family $p(t) = (p_0(t), p_1(t), p_2(t), p_3(t))$ of asymptotic distributions (for sufficiently large n)

As an immediate specification of theorem 1.23, we obtain the main result:

Theorem 4.3 *Let Q denote a periodic $BMAP/PH/c$ queue with period T. Q has a periodic family of asymptotic distributions if and only if the stability condition (4.2) holds. In this case, the periodic family of asymptotic distributions is uniquely determined by*

$$q_s = q\, P_{st}$$

for all $s \in [0, T[$, with q being the stationary distribution of the homogeneous Markov chain $Y = (Y_n : n \in I\!N_0)$ with transition probability matrix $P_{0,T}$.

As intuitively plausible, this theorem shows that periodic Markovian arrival rates yield a periodic asymptotic behaviour of the queue. The stability condition coincides with intuition, too, as it compares the accrued workload and service capacity over one period length. For the special case of $m = 1$ and constant service capacity, the results yield an analysis of the periodic $M_t/M/c$ queue. This extends the results of Heyman and Whitt [68].

Chapter 5

THE $BMAP/G/\infty$ QUEUE

The present chapter gives exact transient distributions as well as approximation formulae for the general (i.e. possibly inhomogeneous) $BMAP/G/\infty$ queue. For special cases of homogeneous $BMAP/G/\infty$ queues, stability conditions and asymptotic distributions are given. Furthermore, it is shown how to derive bounds for the (inhomogeneous) $BMAP/G/c/c$ loss system by means of analyzing the (inhomogeneous) $BMAP/G/\infty$ queue.

The inhomogeneous $BMAP/G/\infty$ queue will be examined using the same method of analysis as the classical one for the $M/G/\infty$ queue (cf. Kalashnikov [73], p.80–82) or the inhomogeneous $M_t/G/\infty$ queue (cf. Eick et al. [54]). Thus many results by Eick et al. [54] for the $M_t/G/\infty$ queue can be proven for the inhomogeneous $BMAP/G/\infty$ queue as well, especially the possibility of using time–dependent service time distributions.

This chapter is organized as follows. The $BMAP/G/\infty$ queue shall be examined in two stages of generality. Essentially, the same method of determining the queue process applies to both of them. In a first step (section 1), this method will be shown for the simplest model of time–homogeneous arrival rates. In this case it is possible to derive statements on stability and asymptotic behaviour, which is done in section 2. Section 3 treats inhomogeneous arrival rates and time–dependent service time distributions. An approximation formula is derived in section 4.

An application of the results to communication systems is given in section 5, which provides bounds for the (inhomogeneous) $BMAP/G/c/c$ loss system. This has the same form as the approximation of multi–server loss systems by infinite server queues announced in Eick et al. [54, 53]. For more application areas of $BMAP/G/\infty$ queues, see the account in Massey and Whitt [84].

1. The Homogeneous Case

In this section, the $BMAP/G/\infty$ queue with homogeneous arrival rates is examined. Let Q denote a queue with a homogeneous BMAP arrival process (N, J) and infinitely many independent servers. Let $\Delta = (D_k : k \in \mathbb{N}_0)$ denote the characteristic sequence determining (N, J). Every incoming user is served immediately, i.e. there is no waiting time and the queue length is always zero. Further, let G denote the service time distribution function.

Define $P_{k;i,j}(s,t)$ as the probability of observing k users being served at time t and the system being in phase $j \in \Phi$ under the condition that at time $s < t$ there were no users being served in S and the system was in phase $i \in \Phi$. Let $P_k(s,t)$ denote the matrix with entries $P_{k;i,j}(s,t)$. Further define $P(s,t) := (P_k(s,t) : k \in \mathbb{N}_0)$ and $P(t) := P(0,t)$.

Fix any time $t \in \mathbb{R}_0^+$ the queue shall be observed at. In any infinitesimal time interval $du = \lim_{h\to 0}]u, u+h]$ with $u \in [0, t[$, batch arrivals occur with rates $\Delta = (D_k : k \in \mathbb{N}_0)$ according to the BMAP arrival process. Given that an arrival of batch size $k \in \mathbb{N}$ occured during $]u, u+h]$, the probability of $n \in \{0, \ldots, k\}$ arrivals still being served at time t is distributed binomially by

$$\binom{k}{n} G^c((t-u)-)^n G((t-u)-)^{k-n},$$

denoting $G((t-u)-) := \lim_{h\to 0} G(t-(u+h))$ and $G^c(t) := 1 - G(t)$.

Conditioning upon the size of the batch arrivals, the total rate of n arrivals during a time interval du which are still in service at time t is

$$R_n(u,t) = \sum_{k=n}^{\infty} D_k \binom{k}{n} G^c((t-u)-)^n G((t-u)-)^{k-n} \tag{5.1}$$

for every $u \in [0, t[$ and $n \in \mathbb{N}_0$.

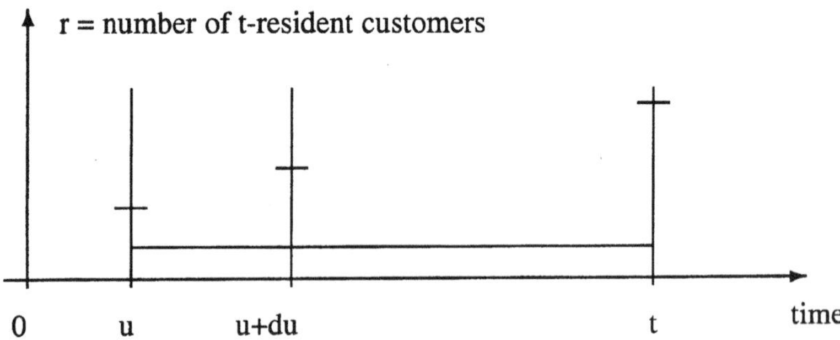

The sequence $R(u,t) := (R_n(u,t) : n \in \mathbb{N}_0)$ can be interpreted as the time–dependent characteristic sequence of an inhomogeneous BMAP $H_t =$

$(H_t(u) : u \leq t)$ which at time $u = t$ coincides in distribution with the infinite server queue that is to be examined. Note that for every time $t \in \mathbb{R}_0^+$ the queue is observed at, the sequence $R(u,t)$ and hence the BMAP H_t is different. The transient distribution of the queue process at time $t \in \mathbb{R}_0^+$ is determined by $P(t) = H_t(t)$.

For two sequences $A = (A_n : n \in \mathbb{N}_0)$ and $B = (B_n : n \in \mathbb{N}_0)$ of $m \times m$ matrices, define the convolution sequence $C = (C_n : n \in \mathbb{N}_0) = A * B$ by $C_n := \sum_{i=0}^{n} A_i B_{n-i}$ for all $n \in \mathbb{N}_0$. Then formula (3.3) yields the following representation:

Theorem 5.1 *The transient distribution of the marginal queue process of Q at time $t \in \mathbb{R}_0^+$ is determined by*

$$P(t) = \sum_{k=0}^{\infty} \underbrace{\int_0^t \int_0^{u_k} \cdots \int_0^{u_2}}_{k \text{ integrals}} R(u_1,t) * \ldots * R(u_k,t)\, du_1 \ldots du_k \tag{5.2}$$

defining the case of zero integrals as the sequence $(I, 0, 0, \ldots)$ with I denoting the identity matrix and 0 the zero matrix.

A great disadvantage in the above formula (5.2) lies in the fact that the sequences $R(u,t)$ need to be determined separately for every time $t \in \mathbb{R}_0^+$ the queue shall be observed at. A closed form for all times of observance is obtained by

Theorem 5.2 *The transient distribution of the marginal queue process of Q at time $t \in \mathbb{R}_0^+$ is determined by*

$$P(t) = \sum_{k=0}^{\infty} \underbrace{\int_0^t \int_0^{u_1} \cdots \int_0^{u_{k-1}}}_{k \text{ integrals}} \tilde{R}(u_1) * \ldots * \tilde{R}(u_k)\, du_k \ldots du_1 \tag{5.3}$$

with $\tilde{R}(u) := R(t-u,t)$ being independent of $t > u$.

Proof: By equation (5.1), the rates

$$\tilde{R}_n(u) := R_n(t-u,t) = \sum_{k=n}^{\infty} D_k \binom{k}{n} G^c(u-)^n G(u-)^{k-n}$$

are independent of $t > u$. Starting from equation (5.2), we have

$$P(t) = \sum_{k=0}^{\infty} \underbrace{\int_0^t \int_0^{u_k} \dots \int_0^{u_2}}_{\text{k integrals}} R(u_1, t) * \dots * R(u_k, t)\, du_1 \dots du_k$$

$$= \sum_{k=0}^{\infty} \int_0^t \int_{u_1}^t \dots \int_{u_{k-1}}^t R(u_1, t) * \dots * R(u_k, t)\, du_k \dots du_1$$

$$= \sum_{k=0}^{\infty} \int_0^t \int_0^{u_1} \dots \int_0^{u_{k-1}} R(t - u_1, t) * \dots * R(t - u_k, t)\, du_k \dots du_1$$

which proves the statement.
☺

The state space of the process Q is $I\!N_0 \times \Phi$, hence the same as that of the BMAP (N, J). Further, the phase process of the queue coincides with that of the BMAP, since the servers do not influence it. Hence we can define an expectation matrix $E(Q_t - Q_s)$ of the queue process Q over the time interval $]s, t]$ exactly as we have defined an expectation kernel for MAJPs, namely by

$$E(Q_t - Q_s)(i, j) := E\left((N_t - N_s) \cdot \delta_{J_t, j} | J_s = i\right)$$

for all $s < t$ and $i, j \leq m$. As a natural abbreviation, we write $E(Q_t) := E(Q_t - Q_0)$ with the understanding that $Q_0 = 0$.

The next two theorems yield expressions for the expectation matrix of the queue process at any time of observance.

Theorem 5.3 *Assume that the arrival rate is finite for any phase $i \in \Phi$, i.e.*

$$\sum_{n=1}^{\infty} \sum_{j=1}^{m} n D_n(i, j) < \infty \tag{5.4}$$

for all $i \in \Phi$. Then the expectation matrix of the queue process Q at time $t \in I\!R_0^+$ is given by

$$E(Q_t) = \int_0^t P_{0u}^{\Phi} \sum_{n=1}^{\infty} n D_n P_{ut}^{\Phi}\, G^c((t - u)-)\, du \tag{5.5}$$

with P_{st}^{Φ} denoting the transition probability matrix of the phase process.

Proof: Fix a time $t \in I\!R_0^+$. Since $P(t)$ equals the distribution of the BMAP H_t at time t, the expectation kernel at time t is given by equation (3.4) as

$$\begin{aligned} E(Q_t) &= \int_0^t P_{0u}^\Phi \sum_{n=1}^\infty n R_n(u,t) P_{ut}^\Phi \, du \\ &= \int_0^t P_{0u}^\Phi \sum_{n=1}^\infty n \sum_{k=n}^\infty D_k \binom{k}{n} p^n(u) q^{k-n}(u) P_{ut}^\Phi \, du \end{aligned}$$

abbreviating $p(u) := G^c((t-u)-)$ and $q(u) := G((t-u)-)$. We further obtain

$$\begin{aligned} E(Q_t) &= \int_0^t P_{0u}^\Phi \sum_{k=1}^\infty \sum_{n=1}^n D_k \frac{k!}{n! \cdot (k-n)!} p^n(u) q^{k-n}(u) P_{ut}^\Phi \, du \\ &= \int_0^t P_{0u}^\Phi \sum_{k-1}^\infty k D_k p(u) \sum_{n=1}^k \frac{(k-1)!}{(n-1)! \cdot (k-n)!} \\ &\quad \times p^{n-1}(u) q^{k-n}(u) P_{ut}^\Phi \, du \end{aligned}$$

which leads to

$$E(Q_t) = \int_0^t P_{0u}^\Phi \sum_{k=1}^\infty k D_k G^c((t-u)-) P_{ut}^\Phi \, du$$

since

$$\begin{aligned} &\sum_{n=1}^k \frac{(k-1)!}{(n-1)! \cdot (k-n)!} p^{n-1}(u) q^{k-n}(u) \\ &\qquad = \sum_{n=0}^{k-1} \frac{(k-1)!}{n! \cdot (k-1-n)!} p^n(u) q^{k-1-n}(u) \end{aligned}$$

sums up to one.
☺

Theorem 5.4 *If condition (5.4) holds and the phase process has stationary distribution π, then the expectation of the number of users in the queue at time $t \in I\!R_0^+$ starting in phase equilibrium is*

$$E_\pi(Q_t) = \pi E(Q_t) 1_m = \pi \sum_{n=1}^\infty n D_n 1_m \cdot \int_0^t G^c(u-) \, du \qquad (5.6)$$

for all $t \in I\!R_0^+$.

Proof: This follows from the above theorem 5.3. Since $\pi P^{\Phi}_{0u} = \pi$ for all $u \in I\!R^+_0$ and $P^{\Phi}_{ut}1_m = 1_m$ for all $u < t \in I\!R^+_0$, we have

$$\begin{aligned}\pi E(Q_t)1_m &= \int_0^t \pi \sum_{n=1}^{\infty} nD_n 1_m G^c((t-u)-)\, du \\ &= \pi \sum_{n=1}^{\infty} nD_n 1_m \cdot \int_0^t G^c(v-)\, dv\end{aligned}$$

after substituting $v := t - u$.
☺

2. Asymptotic Stability

The asymptotic behaviour of the homogeneous BMAP/G/∞ queue already is apparent in formulae (5.3) and (5.5) of the transient distributions and expectation matrices, respectively. The marginal expectation $E(Q_t)$ remains finite for $t \to \infty$ if and only if the mean service time $E(G)$ as well as the mean arrival rate are finite. As will be shown in this section, these conditions are, as intuition would suggest, the main ingredients of a stability condition for the homogeneous SMAP/G/∞ queue.

A homogeneous queue $Q = (Q_t : t \in I\!R^+)$ shall be called **stable** if the asymptotic distribution $p := \lim_{t\to\infty} P(t)$ does exist and the marginal asymptotic distribution of users in the queue has finite expectation.

Theorem 5.5 *Denote the asymptotic distribution of the phase process J by π. If Q has an asymptotic distribution, then the asymptotic expectation of the number of users in the queue is given by*

$$\lim_{t\to\infty} E(Q_t) = E(G) \cdot \pi \sum_{n=1}^{\infty} nD_n 1_m$$

independently of the initial distribution.

Proof: This follows from equation (5.5) if one recognizes that with t growing to infinity, the term $G^c((t-u)-)$ tends to zero for the times u during which the phase process is not in equilibrium yet.
☺

Theorem 5.6 *Let Q denote a homogeneous BMAP/G/∞ queue. Denote the asymptotic distribution of the phase process J by π. Then Q is stable if and only if the stability condition*

$$\pi \sum_{n=1}^{\infty} nD_n 1_m \cdot E(G) < \infty \tag{5.7}$$

holds.

Proof: First we show necessity. Assume that the queue is stable. Then the asymptotic expectation of the marginal process N is given by

$$\lim_{t\to\infty} E(Q_t) = E(G)\cdot\pi\sum_{n=1}^{\infty} nD_n 1_m < \infty$$

according to theorem 5.5. This yields the necessity of condition (5.7).

Sufficiency will be shown by comparison to the $M/G/\infty$ queue. Denote the work load process of Q by $W = (W_t : t \in I\!R_0^+)$. Define $\tilde{Q}$ as the following $M/G/\infty$ queue. The arrival process of $\tilde{Q}$ shall be a Poisson process with rate

$$\gamma_{\max} := -\min_{i\in\Phi} D_0(i,i) = \max_{i\in\Phi}|D_0(i,i)| < \infty$$

Let $i_{\max} \in \Phi$ denote a phase with the highest expectation of the arrival batch size, i.e.

$$\frac{1}{\gamma_{i_{\max}}}\sum_{n=1}^{\infty}\sum_{h=1}^{m} nD_n(i_{\max},h) \geq \frac{1}{\gamma_j}\sum_{n=1}^{\infty}\sum_{h=1}^{m} nD_n(j,h)$$

for all $j \in \Phi$. Since there are no absorbing states, the exit rate $\gamma_{i_{\max}}$ is a positive number. Define the service time distribution of $\tilde{Q}$ by

$$\tilde{G} := \frac{1}{\gamma_{i_{\max}}}\sum_{n=1}^{\infty}\sum_{h=1}^{m} D_n(i_{\max},h)G^{*n}$$

denoting the service time distribution of Q by G and the n–fold convolution of G by G^{*n}. The $M/G/\infty$ queue constructed in this way has an arrival rate which is an upper bound of the phase specific arrival rates of Q. Furthermore, the batch arrivals of Q are interpreted as single arrivals in $\tilde{Q}$ with only one server working off the accrued service requirement of all the users in the batch arrival. Thus the work load $\tilde{W} = (\tilde{W}_t : t \in I\!R_0^+)$ of $\tilde{Q}$ is certainly at least as high as the work load W of Q. Since by Wald's equation

$$\begin{aligned} E(\tilde{G}) &= \frac{1}{\gamma_{i_{\max}}}\sum_{n=1}^{\infty}\sum_{h=1}^{m} D_n(i_{\max},h)\cdot nE(G) \\ &= \frac{1}{\gamma_{i_{\max}}}\cdot E(G)\cdot\sum_{n=1}^{\infty}\sum_{h=1}^{m} nD_n(i_{\max},h) \end{aligned}$$

we have $E(\tilde{G}) < \infty$ by assumption of the stability condition. This implies that $\tilde{W}$ is positive recurrent. i.e. the expected duration between the time instants of $\tilde{W}$ reaching the state 0 is finite. Since

$$\tilde{W}_t \geq W_t \geq 0$$

certainly for all $t \in \mathbb{R}_0^+$, the work load process W of Q is positive recurrent, too. Hence Q has an asymptotic distribution. The finiteness of the asymptotic expectation of the marginal process N is immediate from theorem 5.5 and the stability condition.
☺

For the special case of an $M^B/G/\infty$ queue, i.e. an infinite server queue with general service time distributions and a compound Poisson arrival process, we can state the following results regarding the stationary distribution of the number of users in the system.

Such a queue is specified by setting $D_n := \lambda \cdot p(n)$ for all $n \in \mathbb{N}_0$, with $p(0) = -1$ and $(p(n) : n \in \mathbb{N})$ denoting the probability distribution of the size of a batch arrival. Since the D_n are reduced to scalars in this case, we have commutability of $(\tilde{R}(u) : u \in \mathbb{R}_0^+)$ and hence an exponential form of the transient distribution (see section 4 of chapter 1), which is now given as

$$P(t) = \exp\left(\int_0^t \tilde{R}(u)du\right) := \sum_{k=0}^{\infty} \frac{\left(\int_0^t \tilde{R}(u)du\right)^k}{k!}$$

In particular, we have $\tilde{R}_0(u) = \lambda \cdot \sum_{n=0}^{\infty} p(n)G(u-)^n$ and (assuming stability of the queue) for the asymptotic probability

$$p_0 := \lim_{t\to\infty} P(Q_t = 0) = \lim_{t\to\infty} \exp\left(\int_0^t \lambda \cdot \sum_{n=0}^{\infty} p(n)G(u-)^n du\right)$$
$$= \exp\left(\int_0^{\infty} \lambda \cdot \sum_{n=0}^{\infty} p(n)G(u-)^n du\right)$$

since the sum $\sum_{n=0}^{\infty} p(n)G(u-)^n$ will never be positive and the extended definition $\exp(-\infty) := 0$ is consistent.

For the case $p(1) = 1$ we obtain the classical $M/G/\infty$ queue and the fact that then

$$\tilde{R}_0(u) = -\lambda + \lambda G(u-) = -\lambda \cdot G^c(u-)$$

yields the usual expression

$$p_0 = \exp\left(\int_0^{\infty} -\lambda \cdot G^c(u-)du\right) = \exp\left(-\lambda \cdot E(G)\right)$$

with $E(G)$ denoting the mean service time.

3. The General Case

The infinite server queue fed by an inhomogeneous BMAP can be analyzed analogously to the homogeneous case in section 1. Let $Q = (Q_t : t \in \mathbb{R}_0^+)$

denote a queue with general BMAP input (N, J), general service time distribution G and infinitely many servers. The arrival process (N, J) shall be defined by its time–dependent characteristic sequence $(\Delta(t) : t \in \mathbb{R}_0^+)$.

Be $t \in \mathbb{R}^+$ the time instant the queue is observed at. Now the arrival rates at times $s \in [0, t]$ are not constant anymore, but depend on the time instant s. The total rate of n arrivals during a time interval $du = \lim_{h\to 0}]u, u+h]$ is

$$R_n(u,t) = \sum_{k=n}^{\infty} D_k(u) \binom{k}{n} G^c((t-u)-)^n G((t-u)-)^{k-n}$$

for every $u \in [0, t[$ and $n \in \mathbb{N}_0$. As in the case of homogeneous arrival rates, the sequence $R(u,t) = (R_n(u,t) : n \in \mathbb{N}_0)$ can be interpreted as the time–dependent sequence of transition rates of an inhomogeneous BMAP $H_t = (H_t(u) : u \le t)$ which at time $u = t$ coincides in distribution with the infinite server queue that is to be examined. The same arguments as in section 1 lead to

Theorem 5.7 *The transient distribution of the process Q at time $t \in \mathbb{R}_0^+$ is determined by*

$$P(t) = \sum_{k=0}^{\infty} \underbrace{\int_0^t \int_0^{u_k} \cdots \int_0^{u_2}}_{k \text{ integrals}} R(u_1, t) * \ldots * R(u_k, t)\, du_1 \ldots du_k \tag{5.8}$$

defining the case of zero integrals as the sequence $(I, 0, 0, \ldots)$ with I denoting the identity matrix and 0 the zero matrix, and further defining the convolution sequence $C = A * B$ by $C_n := \sum_{i=0}^{n} A_i B_{n-i}$ for any two sequences A, B of matrices.

Unfortunately, a unification of the above formula analogous to formula (5.3) cannot be achieved in general. This is due to the fact that

$$R_n(t-u,t) = \sum_{k=n}^{\infty} D_k(t-u) \binom{k}{n} G^c(u-)^n G(u-)^{k-n}$$

still depends on t if the arrival rates are inhomogeneous.

On the other hand, it is possible to extend the theory towards time–dependent service time distribution functions. This would lead to rates of the form

$$R_n(u,t) = \sum_{k=n}^{\infty} D_k(u) \binom{k}{n} G_u^c((t-u)-)^n G_u((t-u)-)^{k-n}$$

for every $u \in [0, t[$ and $n \in \mathbb{N}_0$, with G_u denoting the service time distribution function which holds for users arriving during the infinitesimal time interval du.

4. An Approximation

This section shows an approximation method which can be used in order to determine the distribution of the queue process at an arbitrary time $t \in I\!R_0^+$ without needing to integrate over the whole range $[0, t]$. The approximation regards only arrivals up to a certain time distance into the past. All arrivals which have occured before are neglected. Thus, this method approximates the service time distribution by a distribution with cut tail.

Conditions for this approximation are the standard assumptions of finite mean service time and finite arrival rates. First, it is shown that the matrices of the characteristic sequence of H_t are uniformly bounded. Then, this result is used to estimate the approximation error.

Assume in this section that the mean service time

$$E(G) = \int_0^\infty G^c(u)\, du < \infty \tag{5.9}$$

is finite. Further assume that the arrival rate

$$\sum_{n=1}^{\infty} \sum_{j=1}^{m} n D_n(t)(i,j) < M < \infty$$

is bounded by a finite value $M \in I\!R^+$ for all $t \in I\!R^+$ and $i \in \Phi$. The next result gives the decisive bound for the approximation following.

Lemma 5.8 *For every $\varepsilon > 0$, there is a time distance $T(\varepsilon) \in I\!R^+$ such that the inequality*

$$\sup_{1 \le i \le m} \left| \sum_{j=1}^{m} \int_0^s R_n(u,t)(i,j)\, du \right| < M \cdot \varepsilon$$

holds for all $n \in I\!N_0$ and $s \le t - T(\varepsilon)$.

Proof: Choose any $\varepsilon > 0$ and abbreviate $\|L\| := \sup_{i \in \Phi} \sum_{j=1}^{m} L(i,j)$ for matrices L. By assumption (5.9), there is a $T(\varepsilon) \in I\!R^+$ such that

$$\int_{T(\varepsilon)}^{\infty} G^c(u-)\, du < \varepsilon$$

Fix any $y \in \Phi$ and $s \le t - T(\varepsilon)$. Since all elements of $R_n(u,t)$ are positive for all $u \in [0, t]$ and $n \in I\!N$, we have

$$\left\| \int_0^s \sum_{n=1}^{\infty} R_n(u,t)\, du \right\| \le \left\| \int_0^s \sum_{n=1}^{\infty} n R_n(u,t)\, du \right\|$$

Analogously to the proof of theorem 5.3, one can show that

$$\int_0^s \sum_{n=1}^{\infty} nR_n(u,t)\, du = \int_0^s \sum_{n=1}^{\infty} nD_n(u)\; G^c((t-u)-)\, du$$

For all $k \in I\!N$, the estimation

$$\begin{aligned}
\left\| \int_0^s R_k(u,t)\, du \right\| &\leq \left\| \int_0^s \sum_{n=1}^{\infty} R_n(u,t)\, du \right\| \\
&\leq \left\| \int_0^s \sum_{n=1}^{\infty} nD_n(u)\; G^c((t-u)-)\, du \right\| \\
&\leq \int_0^s \left\| \sum_{n=1}^{\infty} nD_n(u) \right\| \; G^c((t-u)-)\, du \\
&< M \cdot \int_0^s G^c((t-u)-)\, du
\end{aligned}$$

holds. Substituting $v := t - u$ leads to

$$\begin{aligned}
\left\| \int_0^s R_k(u,t)\, du \right\| &< M \cdot \int_{t-s}^{t} G^c(v-)\, dv \leq M \cdot \int_{T(\varepsilon)}^{\infty} G^c(v-)\, dv \\
&< M \cdot \varepsilon
\end{aligned}$$

Since

$$\int_0^s \sum_{n=1}^{\infty} \sum_{j=1}^{m} R_n(u,t)(i,j) du = - \int_0^s \sum_{j=1}^{m} R_0(u,t)(i,j) du$$

for all $s \in I\!R_0^+$, the proof is complete.
☺

Using the above bound, we have the following approximation:

Theorem 5.9 *At any time $t \in I\!R^+$ and for every $i \in \Phi$, the distribution of the queue process can be approximated by*

$$\sum_{j=1}^{m} P(t)(i,j) = \sum_{j=1}^{m} P(0,t)(i,j) \approx \sum_{j=1}^{m} P(t-T(\varepsilon),t)(i,j)$$

The approximation error is at most

$$\left| \sum_{j=1}^{m} P_n(t)(i,j) - \sum_{j-1}^{m} P_n(t-T(\varepsilon),t)(i,j) \right| < M \cdot \varepsilon$$

for all $n \in I\!N_0$ and $i \in \Phi$.

Proof: For every $k \in I\!N$, the difference between the respective k-fold integrals appearing in formula (5.8) is

$$\begin{aligned}
&\int_0^t \cdots \int_0^{u_{k-1}} R(u_k,t) * \ldots * R(u_1,t)\, du_k \ldots du_1 \\
&\qquad - \int_{t-T(\varepsilon)}^t \cdots \int_{t-T(\varepsilon)}^{u_{k-1}} R(u_k,t) * \ldots * R(u_1,t)\, du_k \ldots du_1 \\
&= \int_0^t \cdots \int_0^{u_{k-2}} \int_0^{t-T(\varepsilon)} R(u_k,t) * \ldots * R(u_1,t)\, du_k \ldots du_1 \\
&= \int_0^{t-T(\varepsilon)} R(u_k,t)\, du_k \\
&\qquad * \int_0^t \cdots \int_0^{u_{k-2}} R(u_{k-1},t) * \ldots * R(u_1,t)\, du_{k-1} \ldots du_1
\end{aligned}$$

Summing up over all $k \in I\!N_0$ yields

$$\begin{aligned}
Q(t)-&Q(t-T(\varepsilon),t) \\
&= \int_0^{t-T(\varepsilon)} R(u,t)\, du \\
&\qquad * \sum_{k=0}^{\infty} \int_0^t \cdots \int_0^{u_{k-1}} R(u_k,t) * \ldots * R(u_1,t)\, du_k \ldots du_1 \\
&= \int_0^{t-T(\varepsilon)} R(u,t)\, du * P(t)
\end{aligned}$$

Hence, for any $i \in \Phi$ and $n \in I\!N_0$ we have

$$\begin{aligned}
&\left| \sum_{j=1}^{m} P_n(t)(i,j) - \sum_{j=1}^{m} P_n(t-T(\varepsilon),t)(i,j) \right| \\
&\qquad = \left| \sum_{k=0}^{n} \int_{u=0}^{t-T(\varepsilon)} \sum_{h,j=1}^{m} R_k(u)(i,h) du\; P_{n-k}(t)(h,j) \right| \\
&\qquad \leq \max_{k=0,\ldots,n} \left| \int_0^{t-T(\varepsilon)} \sum_{h=1}^{m} R_k(u)(i,h)\, du \right|
\end{aligned}$$

since for every $h \in \Phi$ the probability $\sum_{k=0}^{n} \left| \sum_{j=1}^{m} P_{n-k}(t)(h,j) \right|$ does not exceed one. Because

$$\max_{k=0,\dots,n} \left| \int_0^{t-T(\varepsilon)} \sum_{h=1}^{m} R_k(u)(i,h)du \right| < M \cdot \varepsilon$$

according to theorem 5.8, the proof is complete.
☺

Concluding this section, we give two immediate applications of the above approximation for a computation of the asymptotic distribution of the homogeneous infinite server queue. Let π denote the asymptotic phase distribution of the BMAP. Further, let 1_m denote the m–dimensional column vector with all entries being one.

Corollary 5.10 *Let Q denote a stable homogeneous $BMAP/G/\infty$ queue with service time distribution G. If the service time is bounded, i.e. if there is a time T such that $G(T) = 1$, then the asymptotic distribution of Q is given by*

$$p = \sum_{k=0}^{\infty} \int_0^T \int_0^{u_1} \dots \int_0^{u_{k-1}} \pi \left(\tilde{R}(u_1) * \dots * \tilde{R}(u_k) \right) 1_m \;\; du_k \dots du_1$$

with rate matrices $\tilde{R}(u)$ as defined in section 1.

Corollary 5.11 *Let Q denote a stable homogeneous $SMAP/G/\infty$ queue with service time distribution G. Choose any $\varepsilon > 0$ and determine $T(\varepsilon)$ as in lemma 5.8. Then an approximation of the asymptotic distribution of Q is given by*

$$p \approx \sum_{k=0}^{\infty} \int_0^{T(\varepsilon)} \int_0^{u_1} \dots \int_0^{u_{k-1}} \pi \left(\tilde{R}(u_1) * \dots * \tilde{R}(u_k) \right) 1_m \;\; du_k \dots du_1$$

again with rate kernels $\tilde{R}(u)$ as defined in section 1. The approximation error is uniformly bounded by $M \cdot \varepsilon$, with M denoting a bound for the arrival rate of the BMAP.

5. Bounds for the $BMAP/G/c/c$ Loss System

Now the results which have been obtained shall be applied to support the planning procedure of a communication network. Assume that a network with a maximum loss (or outage) probability of ε is to be designed, which means that a user who wants to use the network shall find it busy with a probability of at most ε. Since a network with finite capacity can be modelled by a (possibly inhomogeneous) $BMAP/G/c/c$ queue, the search is for the minimal sufficient capacity such that the loss probability can be guaranteed to remain smaller than ε.

Then the analysis of infinite server queues can help to derive a capacity function $C_\varepsilon : I\!R^+ \to I\!N_0$ which yields for every time $t \in I\!R^+$ a number $C_\varepsilon(t)$ such that if the network has the capacity of serving $C_\varepsilon(t)$ users at time t, then the quality of service with respect to the given maximum outage probability ε is guaranteed. Hence, in order to build a network that can guarantee the maximum outage probability ε, it suffices to provide service capacity according to the function C_ε.

In order to determine C_ε, one can proceed defining a $BMAP/G/\infty$ queue with appropriate time–dependent arrival rates. Then either the stationary distribution (in the homogeneous case) or the distributions at a given time of day (in the case of periodic arrival rates with period length T) can be represented by a set of functions $P_t : I\!N_0 \to [0,1]$ which yields the probability $P_t(n)$ of n users being served at day time t. For the homogeneous case, we have a function P instead of a set $(P_t : t \in [0,T])$ of functions which depend on the day time.

These are the distributions for an infinite server queue, which means that the probabilities were derived under the assumption that every user can be served. Hence the work load for any real network with some positive outage probability would be lower. It thus suffices to determine the capacity function C_ε for the infinite server queue in order to find a sufficient capacity function for any real network. Since the outage probability ε usually will be chosen very small, the approximation by an infinite server queue seems reasonably sharp.

Now the capacity function for the infinite server queue can easily be determined as

$$C_\varepsilon(t) := \min\left\{ n \in I\!N_0 : \sum_{k=n+1}^{\infty} P_t(k) < \varepsilon \right\}$$

for all $t \in I\!R^+$. This means that the value $C_\varepsilon(t)$ of the capacity function at a time t is given by the lowest number n of users such that the probability of more than n users being served would be smaller than ε for the respective infinite server queue. Since the work load of any real network is lower than the work load for the infinite server queue, the outage probability of a network satisfying this capacity function can be guaranteed to stay smaller than the given threshold ε.

Chapter 6

MODEL FITTING FOR HOMOGENEOUS BMAPS

Although the concept of BMAPs has gained widespread use in stochastic modelling of communication systems and other application areas, there are few statistical methods of parameter estimation proposed yet. A survey of estimation methods is given in Asmussen [4]. His emphasis is on maximum likelihood estimation and its implementation via the EM algorithm. For the Markov Modulated Poisson Process (MMPP), an EM algorithm has been developed by Ryden [114, 115, 116, 117], whereas Asmussen et al. [7] derived a fitting procedure for phase-type distributions via the EM algorithm. An interesting different approach is given by Botta et al. [20] and Chauveau et al. [37]. They introduce the class of generalized hyperexponential distributions and derive statistical fitting methods by means of inversion of transforms.

However, in order to practically use BMAPs for modelling, statistical model fitting from empirical time series is an essential task. The present chapter contains a specification of the classical EM algorithm (see Dempster et al. [50]) for MAPs and BMAPs as well as a performance comparison to the computationally simpler estimation procedure proposed in Breuer and Gilbert [36], or Breuer [30] for a restricted version.

In section 1, the EM algorithm will be specified for MAPs and BMAPs. Section 2 contains a short description of the estimation procedure for BMAPs introduced in Breuer and Gilbert [36]. In section 3 numerical results of the two procedures are compared. In order to understand the details of the estimation procedures to be introduced in the present paper, it is strongly recommended to be familiar with former EM specifications for PH distributions (see Asmussen et al. [7]) and for the MMPP (see Ryden [116]), which is a special case of the MAP.

Assume that the empirical information observed from an input stream into a queueing system consists of the time instants $(T_1, \ldots, T_N)$ of single arrivals (for MAPs) or the time instants $(T_1, \ldots, T_N)$ of arrivals together with their batch sizes $(b_1, \ldots, b_N)$, with N denoting the number of observed arrivals. The task is to find parameters D_0 and D_1 for a MAP (or $D_0, D_1, \ldots, D_M$, with maximal batch size M, for a BMAP) that optimally fit this arrival stream.

In both estimation procedures that will be introduced, we fix the number of phases for the BMAP model to be a known integer $m \geq 2$. Procedures for estimating the number m of phases are discussed in Ryden [117]. A feasible method without a prior estimation of m is proposed in Jewell [72] as follows. Denote the estimators (as given by the EM algorithm and the simpler procedure described below) for an assumed number m_k of phases by $(\hat{D}_0(k), \hat{D}_1(k), \ldots, \hat{D}_M(k))$. Estimating the parameters by the methods given below for increasing m_k and stopping as soon as the likelihood ratio

$$\frac{f(z|\hat{D}_0(k+1), \hat{D}_1(k+1), \ldots, \hat{D}_M(k+1))}{f(z|\hat{D}_0(k), \hat{D}_1(k), \ldots, \hat{D}_M(k))}$$

is smaller than a threshold value $1+\varepsilon$ leads to a reasonable model fitting. Since the adaptation of the model increases with the assumed number of phases, the likelihood gain is always positive. The threshold value reflects the limit of accuracy beyond which the gain in model adaptation is not worth the additional computation time.

This method has been applied to BMAP estimation via the procedure described in section 2 by Breuer and Gilbert [36]. Many numerical results can be found in Gilbert [63]. In particular, it can be seen there that the number of phases of the estimated BMAP representation does not always coincide with the number of phases of the input BMAP. Nevertheless, in these cases the likelihood of the arrival stream under the estimated representation is about the same as the likelihood under the original input BMAP.

1. An EM algorithm for MAPs and BMAPs

The typical property of observing time series derived from a MAP (resp. a BMAP) is that only the time instants of arrivals (and their batch sizes, resp.) but not the phases can be seen. If the phases were observable, then one could apply classical estimators for finite state Markov processes (see Albert [1] or Basawa and Rao [14]) and the problem would be solved. Thus we have a problem of estimation from incomplete data. For this type of statistical problems, the so–called EM algorithm has proven to be a good means of approximating the maximum likelihood estimator (see Dempster et al. [50], McLachlan and Krishnan [85] or Meng and Dyk [88]). The name EM algorithm stems from the alternating application of an expectation step (for E) and a maximization step (for M) which yield successively higher likelihoods of the estimated parameters.

In this section we first describe the classical estimators which were applicable if we had the complete sample for a MAP (i.e. if phases were observable). Then we derive the specification of the EM algorithm for MAPs, assuming that phases are not observable (incomplete sample). Finally, this specification is extended to obtain an EM specification for BMAPs.

1.1 Complete sample case for MAPs

A sufficient statistic for the complete data sample would be the collection

$$\left(J_n^k : 0 \leq n \leq M_k, 1 \leq k \leq N\right), \left(S_n^k : 0 \leq n \leq M_k - 1, 1 \leq k \leq N\right)$$

of random variables, where J_n^k denotes the phase immediately after the nth jump of the phase process in the interval $[T_{k-1}, T_k[$, M_k the number of jumps in this interval (not including the last jump accompanied by an arrival), and S_n^k the sojourn time after the nth jump of the phase process in the interval $[T_{k-1}, T_k[$.

Then the density of a complete sample x under parameters $D_0 = (D_{0;ij})$ and $D_1 = (D_{1;ij})$ would be given by

$$f(x|D_0, D_1) = \prod_{i=1}^{m} \exp\left(D_{0;ii} Z_i\right) \prod_{i=1}^{m} \prod_{j=1, j\neq i}^{m} D_{0;ij}^{N_{ij}} \prod_{i=1}^{m}\prod_{j=1}^{m} D_{1;ij}^{L_{ij}}$$

where Z_i denotes the total time spent in phase i, N_{ij} the number of jumps from phase i to phase j without accompanying arrival, and L_{ij} the number of jumps from phase i to phase j with accompanying arrival. These variables can be computed from the sufficient statistic by

$$Z_i = \sum_{k=1}^{N} \sum_{n=0}^{M_k - 1} 1_{(J_n^k = i)} S_n^k, \qquad N_{ij} = \sum_{k=1}^{N} \sum_{n=0}^{M_k - 1} 1_{(J_n^k = i)} 1_{(J_{n+1}^k = j)}$$

and

$$L_{ij} = \sum_{k=1}^{N-1} 1_{(J_{M_k}^k = i)} 1_{(J_0^{k+1} = j)}$$

for $1 \leq i \neq j \leq m$. Further define

$$Y_i = \sum_{k=1}^{N-1} \sum_{n=0}^{M_k - 1} 1_{(J_n^k = i)} S_n^k$$

for $1 \leq i \leq m$.

Acknowledging the relation $D_{0;ii} = -\left(\sum_{j=1}^{m} D_{1;ij} + \sum_{j=1,j\neq i}^{m} D_{0;ij}\right)$, the maximum likelihood estimators $\hat{D}_0$ and $\hat{D}_1$ for the matrices D_0 and D_1 would be

$$\hat{D}_{0;ij} = \frac{N_{ij}}{Z_i}, \qquad \hat{D}_{1;ij} = \frac{L_{ij}}{Y_i}, \tag{6.1}$$

$$\hat{D}_{0;ii} = -\left(\sum_{j=1}^{m} \hat{D}_{1;ij} + \sum_{j=1,j\neq i}^{m} \hat{D}_{0;ij}\right) \tag{6.2}$$

for $1 \leq i, j \leq m$, as given in Albert [1].

1.2 EM for MAPs

In the case of observing only an incomplete sample z, the EM algorithm provides an iteration of alternating expectation (E) and maximization (M) steps that lead to a reevaluation of the estimators increasing their likelihoods in every cycle of E- and M–step. In our case, the incomplete sample y consists only of the sequence $(T_0 = 0, T_1, \dots, T_N)$ of arrival times that are observable. Keeping in mind $T_0 = 0$, we will not lose information by setting $z = (z_1, \dots, z_N) := (T_1, T_2 - T_1, \dots, T_n - T_{N-1})$.

Given the parameters D_0 and D_1 as well as an initial phase distribution π, the likelihood of the incomplete sample z is

$$f(z|\pi, D_0, D_1) = \pi \left(\prod_{n=1}^{N-1} \exp(D_0 z_n) D_1\right) \exp(D_0 z_N)\eta \tag{6.3}$$

with $\eta := D_1 1_m$, denoting by 1_m the m–dimensional column vector with all entries being 1.

Assume that the estimates after the kth EM iteration are given by the matrices $(\hat{D}_0^{(k)}, \hat{D}_1^{(k)})$. Then in the first step of the $k+1$st cycle, the conditional expectations of the variables Z_i, N_{ij} and L_{ij} given the incomplete observation y and the current estimates $(\hat{D}_0^{(k)}, \hat{D}_1^{(k)})$ are computed.

In order to simplify notations, define

$$\eta_N(i) := \sum_{j=1}^{m} \hat{D}_1^{(k)}(i,j) \qquad \text{and} \qquad \eta_{n-1}(i) := \hat{D}_1^{(k)} \exp(\hat{D}_0^{(k)} z_n)\eta_n \tag{6.4}$$

for $2 \leq n \leq N$ and $i = 1, \dots, m$.

Since the empirical time series is observed in a stationary regime, we can set the phase distribution π_0 at time $T_0 = 0$ to be the phase equilibrium, i.e. satisfying $\pi_0(\hat{D}_0^{(k)} + \hat{D}_1^{(k)}) = 0$. Thus π_0 is a deterministic function of $(\hat{D}_0^{(k)}, \hat{D}_1^{(k)})$

and in particular $f(z|\pi_0, \hat{D}_0^{(k)}, \hat{D}_1^{(k)}) = f(z|\hat{D}_0^{(k)}, \hat{D}_1^{(k)})$ holds. Next we define iteratively

$$\pi_{n+1} := \pi_n \exp(\hat{D}_0^{(k)} z_{n+1})\, \hat{D}_1^{(k)} \tag{6.5}$$

for $0 \leq n \leq N-2$, interpreting the π_n as row vectors.

We can continue the E–step with

$$Z_i^{(k+1)} := E_{(\hat{D}_0^{(k)}, \hat{D}_1^{(k)})}(Z_i|z) = \sum_{n=1}^{N} E_{(\hat{D}_0^{(k)}, \hat{D}_1^{(k)})}(Z_i^n|z)$$

and

$$Y_i^{(k+1)} := E_{(\hat{D}_0^{(k)}, \hat{D}_1^{(k)})}(Y_i|z) = \sum_{n=1}^{N-1} E_{(\hat{D}_0^{(k)}, \hat{D}_1^{(k)})}(Z_i^n|z)$$

where Z_i^n denotes the random variable of the total amount of time within $[T_{n-1}, T_n[$ that is spent in phase i. This is given by

$$E_{(\hat{D}_0^{(k)}, \hat{D}_1^{(k)})}(Z_i^n|z) = \frac{c_n(i, i|z, \hat{D}_0^{(k)}, \hat{D}_1^{(k)})}{f(z|\hat{D}_0^{(k)}, \hat{D}_1^{(k)})} \tag{6.6}$$

for all $1 \leq l \leq N$, with c_n given as in definition (6.7) below. The derivation of (6.6) is completely analogous to the one in Asmussen et al. [7], p.439. Likewise, the E–step for

$$N_{ij}^{(k+1)} := E_{(\hat{D}_0^{(k)}, \hat{D}_1^{(k)})}(N_{ij}|z) = \sum_{n=1}^{N} E_{(\hat{D}_0^{(k)}, \hat{D}_1^{(k)})}(N_{ij}^n|z)$$

with

$$E_{(\hat{D}_0^{(k)}, \hat{D}_1^{(k)})}(N_{ij}^n|z) = \frac{\hat{D}_{0;ij}^{(k)} c_n(i, j|z, \hat{D}_0^{(k)}, \hat{D}_1^{(k)})}{f(z|\hat{D}_0^{(k)}, \hat{D}_1^{(k)})}$$

for $1 \leq n \leq N$ is derived using completely the same arguments as in Asmussen et al. [7], p.440. Here, the matrix functions c_n are defined as

$$c_n(i, j|z, \hat{D}_0^{(k)}, \hat{D}_1^{(k)}) := \int_0^{z_n} \pi_{n-1} \exp(D_0^{(k)} u) e_i \cdot e_j^T \exp(D_0^{(k)}(z-u)) \eta_n \, du \tag{6.7}$$

for $1 \leq n \leq N$ and $1 \leq i, j \leq m$.

The E–step is completed by

$$L_{ij}^{(k+1)} := E_{(\hat{D}_0^{(k)},\hat{D}_1^{(k)})}(L_{ij}|z) = \sum_{n=1}^{N-1} E_{(\hat{D}_0^{(k)},\hat{D}_1^{(k)})}(L_{ij}^n|z)$$

with

$$\begin{aligned} E_{(\hat{D}_0^{(k)},\hat{D}_1^{(k)})}(L_{ij}^n|z) &= P(J_{M_n}^n = i, J_0^{n+1} = j|z_1,\ldots,z_n,z_{n+1},\ldots,z_N) \\ &= \frac{1}{f(z)} f(J_{M_n}^n = i, z_1,\ldots,z_n) P(J_0^{n+1} = j|J_{M_n}^n = i) \\ &\quad \times f(z_{n+1},\ldots,z_N|J_0^{n+1} = j) \\ &= \frac{\left(\pi_{n-1}\exp(D_0^{(k)}z_n)\right)_i \hat{D}_1^{(k)}(i,j)\left(\exp(D_0^{(k)}z_{n+1})\eta_{n+1}\right)_j}{f(z|\hat{D}_0^{(k)},\hat{D}_1^{(k)})} \end{aligned}$$

for $1 \le n \le N-1$.

The second step of the $k+1$st cycle consists of the computation of maximum likelihood estimates given the new (conditional but complete) statistic computed in the E–step. This can be done by simply replacing the variables in equations (6.1) and (6.2) by the conditional expectations computed above. This leads to reevaluated estimates

$$\hat{D}_{0;ij}^{(k+1)} = \frac{N_{ij}^{(k+1)}}{Z_i^{(k+1)}}, \qquad \hat{D}_{1;ij}^{(k+1)} = \frac{L_{ij}^{(k+1)}}{Y_i^{(k+1)}},$$

and

$$\hat{D}_{0;ii}^{(k+1)} = -\left(\sum_{j=1}^{m} \hat{D}_{1;ij}^{(k+1)} + \sum_{j=1,j\neq i}^{m} \hat{D}_{0;ij}^{(k+1)}\right)$$

for $1 \le i,j \le m$.

Using these, one can compute the likelihood $f(z|\hat{D}_0^{(k+1)},\hat{D}_1^{(k+1)})$ of the empirical time series under the new estimates according to equation (6.3). If the likelihood ratio

$$\rho = \frac{f(z|\hat{D}_0^{(k+1)},\hat{D}_1^{(k+1)})}{f(z|\hat{D}_0^{(k)},\hat{D}_1^{(k)})}$$

remains smaller than a threshold $1+\varepsilon$, then the EM iteration process can be stopped, and the latest estimates may be adopted. The threshold value reflects the limit of accuracy beyond which the gain in model adaptation is considered not to be worth the additional computation time.

1.3 EM for BMAPs

Once we have obtained the above EM specification for MAPs, an extension to BMAPs is straightforward. In difference to MAPs, in the BMAP case we not only know the times $(T_1, \ldots, T_N)$ of arrivals but also the arrival sizes $(b_1, \ldots, b_N)$, meaning that at time instant T_n there was a batch arrival of size b_n, with $n = 1, \ldots, N$, $b_n \in I\!N$. In order to obtain reasonable estimates, we need an upper bound M such that $D_n = 0$ for all $n \geq M$. Furthermore, we need enough arrivals to ensure that there are reasonably many arrival events of every size in the time series.

Starting from the kth estimates $(\hat{D}_0^{(k)}, \hat{D}_1^{(k)}, \ldots, \hat{D}_M^{(k)})$ for $(D_0, \ldots, D_M)$, the $k+1$st EM iteration proceeds completely analogous as for MAPs, with obvious adaptations in equations (6.3), (6.4) and (6.5). The only more substantial difference is that for BMAPs we need to compute $L_{1;ij}^{(k+1)}, \ldots, L_{M;ij}^{(k+1)}$ instead of only $L_{ij}^{(k+1)}$. These are given similarly by

$$L_{s;ij}^{(k+1)} := E_{(\hat{D}_0^{(k)},\ldots,\hat{D}_M^{(k)})}(L_{s;ij}|z) = \sum_{n=1,b_n=s}^{N-1} E_{(\hat{D}_0^{(k)},\ldots,\hat{D}_M^{(k)})}(L_{s;ij}^n|z)$$

for $s = 1, \ldots, M$, with

$$E_{(\hat{D}_0^{(k)},\ldots,\hat{D}_M^{(k)})}(L_{s;ij}^n|z) = \frac{\left(\pi_{n-1}\exp(D_0^{(k)} z_n)\right)_i \hat{D}_s^{(k)}(i,j)\left(\exp(D_0^{(k)} z_{n+1})\eta_{n+1}\right)_j}{f(z|\hat{D}_0^{(k)}, \hat{D}_1^{(k)}, \ldots, \hat{D}_M^{(k)})}$$

for $1 \leq n \leq N-1$.

These expectations lead to new estimates

$$\hat{D}_{s;ij}^{(k+1)} = \frac{L_{s;ij}^{(k+1)}}{Y_i^{(k+1)}}$$

for $s = 1, \ldots, M$.

2. A simpler estimation procedure

In Breuer and Gilbert [36], a computationally much lighter estimation procedure has been introduced. It works in three steps, each of which uses classical statistical methods. In the first step (section 2.1), the empirical interarrival times are used to estimate the matrix D_0, neglecting the correlations between consecutive interarrival times. Then those can be interpreted as a sample of phase type distributions and hence the entries of D_0 can be estimated by an

EM–algorithm for phase–type distributions (see Asmussen et al. [7]). In the second step (section 2.2), for every empirical arrival instant the probability distribution of being in a certain phase immediately before resp. after this instant is estimated using discriminant analysis (see Titterington et al. [122]). In the last step (section 2.3), the derived estimators of the first two steps are used in order to calculate the empirical estimator for the matrices D_n, $n \geq 1$. This is done according to standard estimators for Markov chains (see Anderson and Goodman [2]).

2.1 Estimating the Matrix D_0

As in section 1, let $(z_n := T_n - T_{n-1} : n \in \{1, \ldots, N\})$ denote the empirical interarrival times. Denote the number of phases by m. According to Baum [16], p.42, the interarrival times of a BMAP are distributed phase–type with generator D_0. Hence if we first neglect the correlations between consecutive inter–arrival times, then we can interpret the $(z_n : n \in \{1, \ldots, N\})$ as a sample of a phase–type distribution with density

$$z(t) = \pi e^{D_0 t} \eta$$

for $t \in I\!R_+$. Here, $\pi = (\pi_1, \ldots, \pi_m)$ is the constant distribution of the phase process immediately after arrival instants and $\eta := -D_0 1_m$ is the so–called exit vector of the phase–type distribution with representation (π, D_0).

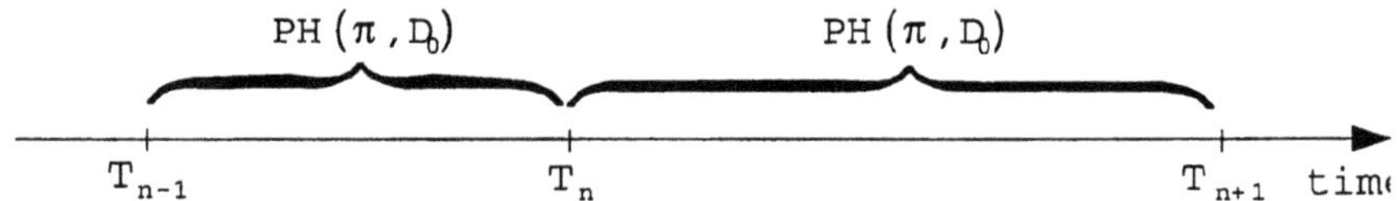

Neglecting the correlations between inter–arrival times

If m is known, there is a maximum likelihood estimator for π and D_0. The solution of the estimating equations can be approximated iteratively by an EM algorithm (cf. Dempster et al. [50] or McLachlan and Krishnan [85]), which was derived for this special case by Asmussen et al. [7] and proceeds as follows.

Starting from an intuitive first estimation $(\pi^{(1)}, D_0^{(1)})$ of the representation of the phase–type distribution, the recursions

$$\pi_i^{(k+1)} = \frac{1}{N} \sum_{n=1}^{N} \frac{\pi_i^{(k)} b_i^{(k)}(z_n)}{\pi^{(k)} b^{(k)}(z_n)}$$

$$D_{0;ij}^{(k+1)} = \sum_{n=1}^{N} \frac{D_{0;ij}^{(k)} c_{ij}^{(k)}(z_n)}{\pi^{(k)} b^{(k)}(z_n)} \Bigg/ \sum_{i=1}^{N} \frac{c_{ii}^{(k)}(z_n)}{\pi^{(k)} b^{(k)}(z_n)}$$

for $i \neq j$ and

$$\eta_i^{(k+1)} = \sum_{n=1}^{N} \frac{\eta_i^{(k)} a_i^{(k)}(z_n)}{\pi^{(k)} b^{(k)}(z_n)} \Big/ \sum_{i=1}^{N} \frac{c_{ii}^{(k)}(z_n)}{\pi^{(k)} b^{(k)}(z_n)}$$

along with the relation

$$D_{0;ii}^{(k+1)} = -\eta_i^{(k+1)} - \sum_{j=1, j\neq i}^{m} D_{0;ij}^{(k+1)}$$

and the definitions

$$a^{(k)}(z_n) := \pi^{(k)} e^{D_0^{(k)} z_n}$$
$$b^{(k)}(z_n) := e^{D_0^{(k)} z_n} \eta^{(k)}$$
$$c_{ij}^{(k)}(z_n) := \int_0^{z_n} \pi^{(k)} e^{D_0^{(k)} u} c_i c_j^T e^{D_0^{(k)}(z_n - u)} \eta^{(k)} \, du$$

for $i, j \in \{1, \ldots, m\}$ und $k \in I\!N$ lead to monotonically increasing likelihoods.

In Asmussen et al. [7], it is proposed to compute the values of $a^{(k)}(z_n)$, $b^{(k)}(z_n)$ and $c_{ij}^{(k)}(z_n)$ numerically as the solution to a linear system of homogeneous differential equations.

2.2 Phases at Arrival Instants

Using the estimator $(\hat{\pi}, \hat{D}_0)$ from the last section, the distribution of the non–observable phases at times $(T_n : n \in I\!N)$ can be estimated using discriminant analysis in a standard way (cf. Titterington et al. [122], pp.168f).

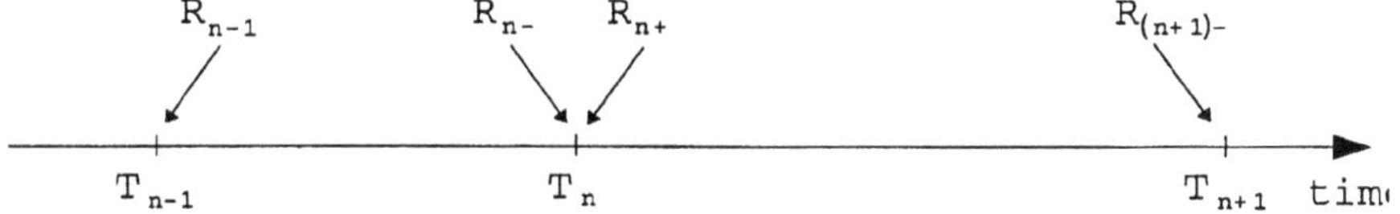

Re–introducing correlations via estimation of phases

For a given empirical arrival instant T_n, let $Z_n := T_n - T_{n-1}$ and R_{n-1} denote the random variables of the last interarrival time and of the phase immediately after the last arrival instant T_{n-1}, respectively. Let $P(Z_n, R_{n-1})$ denote their common distribution and $P(Z_n|R_{n-1})$, $P(R_{n-1}|Z_n)$ the conditional distributions.

Since $(Z_n : n \in \{1, \ldots, N\})$ is given empirically by the time series $(T_n : n \in \{0, \ldots, N\})$, it suffices to estimate the distribution $P(R_{n-1}|Z_n = z_n)$ for every $n \in \{1, \ldots, N\}$. This distribution is discrete, because by assumption there are only finitely many phases.

Bayes' formula yields for $j \in \{1, \ldots, m\}$ and $n \in \{1, \ldots, N\}$

$$P(R_{n-1} = j | Z_n = z_n) = \frac{P(Z_n = z_n | R_{n-1} = j) \cdot P(R_{n-1} = j)}{\sum_{i=1}^{m} P(Z_n = z_n | R_{n-1} = i) \cdot P(R_{n-1} = i)}$$

The expressions $P(Z_n | R_{n-1})$ exist as conditional densities (with respect to the Lebesgue measure on $I\!R$), since the interarrival times Z_n are distributed phase–type.

The expressions on the right hand can be estimated via the estimated parameters $(\hat{\pi}, \hat{D}_0)$ and the resulting vector $\hat{\eta}$. Hence, for every arrival instant T_{n-1} the estimator for the conditional distribution of the phase at T_{n-1} given the interarrival time $Z_n = T_n - T_{n-1}$ is given by

$$\hat{P}(R_{n-1} = i | Z_n = z_n) = \frac{e_i^T e^{\hat{D}_0 z_n} \hat{\eta} \cdot \hat{\pi}_i}{\sum_{j=1}^{m} e_j^T e^{\hat{D}_0 z_n} \hat{\eta} \cdot \hat{\pi}_j} = \frac{\hat{\pi}_i \cdot e_i^T e^{\hat{D}_0 z_n} \hat{\eta}}{\hat{\pi} e^{\hat{D}_0 z_n} \hat{\eta}}$$

for every $i \in \{1, \ldots, m\}$ and $n \in \{1, \ldots, N\}$.

Furthermore, it will be necessary to estimate the phase immediately before an arrival instant. This can be done by the same method. Denote the respective conditional distribution by $P(R_{n-} | Z_n)$. Then we get the estimation

$$\hat{P}(R_{n-} = i | Z_n = z_n) = \frac{\hat{\pi} e^{\hat{D}_0 z_n} e_i \cdot \hat{\eta}_i}{\sum_{j=1}^{m} \hat{\pi} e^{\hat{D}_0 z_n} e_j \cdot \hat{\eta}_j} = \frac{\hat{\pi} e^{\hat{D}_0 z_n} e_i \cdot \hat{\eta}_i}{\hat{\pi} e^{\hat{D}_0 z_n} \hat{\eta}}$$

for every $i \in \{1, \ldots, m\}$ and $n \in \{1, \ldots, N\}$.

2.3 Estimating the Matrices D_n for $n \geq 1$

The matrices $(D_n : n \in I\!N_0)$, which are the blocks of the generator matrix Q of a BMAP of order m, have dimension $(m \times m)$ and the form

$$D_n(i, j) = -D_0(i, i) \cdot p_i(n, j)$$

for every $i, j \in \{1, \ldots, m\}$.

In order to complete the estimation of the generator matrix, it suffices to estimate the parameters $(p_i(n, j) : n \in I\!N, i, j \in \{1, \ldots, m\})$, which are the transition probabilities of the embedded Markov chain at arrival instants. Without any further assumptions regarding these, use of the empirical estimator is standard. This is given in Anderson and Goodman [2]. Since in the present statistical model the phase process is hidden, the phase change at each arrival instant cannot be observed but must be estimated. For this, the results of the last section are used.

For every $k \in \{0, \ldots, N-1\}$, let R_k and R_{k-} denote the random variable of the (non-observable) phase at the empirical arrival instant T_k and immediately before it, respectively. In the last section, estimators for the conditional distributions of the R_k and R_{k-} were given. Let δ denote the Kronecker function and remember that b_k denotes the size of the kth batch arrival. Define

$$n_i(n,j) := \sum_{k=1}^{N-1} \hat{P}(R_{k-} = i | Z_k = z_k) \cdot \delta_{b_k,n} \cdot \hat{P}(R_k = j | Z_{k+1} = z_{k+1})$$

and

$$n_i := \sum_{k=1}^{N-1} \hat{P}(R_{k-} = i | Z_k = z_k) = \sum_{n=1}^{\infty} \sum_{j=1}^{m} n_i(n,j)$$

for $n \in I\!N$ and $i, j \in \{1, \ldots, m\}$. In a BMAP, the random variables R_{k-} and R_k are dependent in any non–trivial case. Since in the present model the phases are non–observable, this dependence can only be reflected by conditioning on the consecutive empirical interarrival times z_k and z_{k+1}.

Because of

$$p_i(n,j) = \frac{P(R_{k-} = i, b_k = n, R_k = j)}{P(R_{k-} = i, b_k \geq 1)} \cdot \frac{P(R_{k-} = i, b_k \geq 1)}{P(R_{k-} = i)}$$

for all $k \in \{1, \ldots, N-1\}$, the empirical estimator for $p_i(n,j)$ is given by

$$\hat{p}_i(n,j) := \frac{n_i(n,j)}{n_i} \cdot \frac{\hat{\eta}_i}{-\hat{D}_{0,ii}}$$

for every $n \in I\!N$ and $i, j \in \{1, \ldots, m\}$.

3. Numerical results and comparison of the two procedures

The following computations were done on a 500 MHz processor using the programming language Octave (a free MatLab version) under Linux. Tables 1 through 3 are results from time series of 100 arrivals, while tables 4 through 6 give estimates from 500 arrivals. The numerical examples were only computed for MAPs, since the estimator for BMAPs is essentially the same but BMAPs would require much larger time series and thus longer run times. The run times for the EM algorithm are further the main reason of not testing time series with more arrivals.

At first glance it can be observed that higher likelihoods do not necessarily mean that the estimated parameters are closer to the original parameters. As a maximum likelihood estimator, the EM algorithm simply tries to find any set of parameters which yields the highest likelihood of the observed time series. The numerical results show that the simpler estimator achieves almost the same likelihood as the EM algorithm while performing much faster and requiring less storage. Since phase–type distributions (and thus BMAPs) in general have no unique representation (see e.g. Botta et al. [20]), it is not reasonable to search for such one. The aim of model fitting is simply to find parameters for a BMAP that make the observed time series under the BMAP model at least as likely as under the original parameters. In all numerical examples with time series derived from MAPs, this happened to be the case for both procedures.

Since MAPs are extensively used as models for arrival streams in queueing systems, a more reasonable measure for the goodness of fit of the tested estimation procedures might be a collection of performance measures for queues with the respective inputs. One such performance measure is the so–called caudal characteristic η (see Neuts [97] or Latouche and Ramaswami [77]). The value of η is a good description of the tail behaviour of the stationary distribution of the queue process.

In the following numerical examples, for every arrival process (as well as for the estimators) the caudal characteristic η of the respective $MAP/M/1$ queue has been computed. This has been computed as the largest eigenvalue of the rate matrix R of the QBD Markov chain embedded at jump times of the queue (including phase changes). The rate of the exponential service time distribution has been set to equal the highest arrival rate (i.e. the absolute value of the lowest entry of the input MAP's D_0). The numerical results show that the caudal characteristics computed from the two estimators are very close and (except in tables 1 and 4) approximate the caudal characteristic of the input MAP well enough.

The EM algorithm yields consistently higher likelihoods than the simpler procedure. Yet, comparing the gain in likelihood per arrival and the total run time, the question remains whether an EM procedure is worth the effort. Although the likelihoods of the estimated parameters are satisfying for both procedures, none of them can approach the original parameters in terms of a maximal distance norm. This can be seen best from the estimated values for D_1. However, in terms of the caudal characteristic for the respective queues with exponential service time distribution, the goodness of fit is much better. The EM algorithm uses all the available information in all iterations. Any improvement of the procedure introduced in section 2 will perform at most as good as the EM algorithm.

Table 1	Input MAP		Estimation by EM		Simple estimation	
$D_0 =$	-10.0	5.0	-10.5	5.4	-15.3	10.1
	50.0	-100.0	30.4	-78.2	59.4	-104.2
$D_1 =$	4.0	1.0	2.8	2.2	2.6	2.7
	10.0	40.0	26.7	21.1	22.1	22.7
run time:			6m 27s		1m 4s	
likelihood:	3.58 E58		1.62 E59		9.86 E58	
η	0.2445		0.1777		0.1752	

Table 2	Input MAP		Estimation by EM		Simple estimation	
$D_0 =$	-5.0	2.0	-3.5	1.4	-5.0	2.5
	5.0	-10.0	8.0	-21.4	5.1	-10.3
$D_1 =$	2.0	1.0	1.8	0.4	1.7	0.9
	3.0	2.0	10.1	3.3	3.4	1.8
run time:			1m 9s		8s	
likelihood:	9.08 E05		5.74 E06		1.51 E06	
η	0.3611		0.3583		0.3432	

Tab 3	Input MAP			Estimation by EM			Simple estimation		
$D_0 =$	-5.0	1.0	1.0	-7.8	2.7	2.6	-11.9	4.0	4.0
	2.0	-10.0	1.0	4.6	-14.7	4.3	6.2	-18.0	6.0
	10.0	2.0	-20.0	6.7	6.3	-23.7	8.7	8.4	-25.5
$D_1 =$	2.0	1.0	0.0	0.9	0.8	0.8	1.7	1.2	0.9
	2.0	4.0	1.0	2.0	1.8	2.0	2.6	1.8	1.4
	1.0	5.0	2.0	3.6	3.3	3.7	3.8	2.6	2.0
time:				27m 56s			14s		
likel.:	1.51 E26			3.43 E26			1.28 E26		
η	0.2627			0.2744			0.2684		

Table 4	Input MAP		Estimation by EM		Simple estimation	
$D_0 =$	-10.0	5.0	-9.5	4.9	-11.2	7.0
	50.0	-100.0	55.9	-119.9	67.2	-123.6
$D_1 =$	4.0	1.0	3.3	1.3	2.2	2.1
	10.0	40.0	28.3	35.6	28.4	28.0
run time:			27m 12s		4m 14s	
likelihood:	1.10 E269		1.17 E270		2.85 E269	
η	0.2445		0.2096		0.1816	

Table 5	Input MAP		Estimation by EM		Simple estimation	
$D_0 =$	-5.0	2.0	-5.4	2.6	-5.6	2.7
	5.0	-10.0	4.8	-10.2	5.0	-10.4
$D_1 =$	2.0	1.0	1.8	0.9	1.9	1.0
	3.0	2.0	3.5	1.8	3.5	1.8
run time:			11m 43s		26s	
likelihood:	3.59 E59		3.98 E59		3.91 E59	
η	0.3611		0.3456		0.3639	

Tab 6	Input MAP			Estimation by EM			Simple estimation		
$D_0 =$	-5.0	1.0	1.0	-5.7	1.8	1.9	-7.0	2.4	2.4
	2.0	-10.0	1.0	4.8	-17.8	5.1	7.0	-21.2	6.1
	10.0	2.0	-20.0	6.9	6.8	-24.6	10.3	8.7	-30.7
$D_1 =$	2.0	1.0	0.0	0.7	0.8	0.6	0.9	0.8	0.6
	2.0	4.0	1.0	2.8	3.0	2.2	3.2	2.9	2.1
	1.0	5.0	2.0	4.0	4.1	2.9	4.6	4.2	3.0
time:				51m 42s			2m 36s		
likel.:	8.53 E117			4.90 E118			1.35 E118		
η	0.2627			0.2869			0.2845		

III

SPATIAL QUEUES

Chapter 7

SPATIAL MARKOVIAN ARRIVAL PROCESSES

The recent rise of mobile communication systems gave reason to the development of queueing models which are capable of incorporating spatial features. Introducing spatial arrival processes is of great help for modelling mobile communication networks, since users of such networks can be characterized more adequately by properties depending on their position. To this end, the concept of a time–space process is needed, which is able to describe the time dynamics of arrivals as well as their distribution in space.

In this chapter, the concept of Batch Markovian Arrival Processes (or shortly BMAPs, see chapter 3) will be generalized towards a class of such time–space processes, which shall be called Spatial Markovian Arrival Processes (or shortly SMAPs). This generalization points in three directions. First, the phase space is allowed to be general. Second, the generator of an SMAP may depend on time, i.e. SMAPs provide models for time–inhomogeneous behaviour. And finally, the attribute "spatial" appearing in the name stems from the feature that arrivals may assume a location in some space. Thus they are distributed not only with respect to time but also with respect to their location in a so–called arrival space. This is important for modelling mobile communication networks, for which the spatial distribution of users is of relevance and often the arrival intensities and service requirements may depend on the location a user calls from. Furthermore, spatial distributions of users typically will include spatial correlations among the different locations of arrivals in a batch.

The spatial features of SMAPs are inspired by and based upon an earlier concept of Markovian spatial arrival processes introduced by Baum [17]. This conception of SMAPs keeps the most important structural properties of BMAPs, such that the computations necessary for analyzing them and the respective queues are still tractable.

Before introducing spatial arrival processes, it is helpful to become more acquainted with concepts of spatial arrivals. These are introduced via the concept of counting functions in section 1. In section 2, the class of spatial Markovian arrival processes is defined and constructed. The most important properties of SMAPs are derived in sections 3 and 4. In particular, the transition probabilities of SMAPs are specified and a law of large numbers is given for periodic SMAPs. The last section contains some simple examples.

1. Stochastic Point Fields

Stochastic point fields can be regarded as stochastic point processes with parameter sets not necessarily being the time axis (as the word "process" hints at) but a general space. They are a very useful concept in many application fields such as astronomy (Babu and Feigelson [9]), ecology (Thompson [120]), the modelling of populations (Moyal [91]), rain fall (Cowpertwait [46], Phelan [106]) and mobile communication networks (Baum [17]). An extensive list of references is given in Cressie [47]. Although in the rest of this book the focus shall rest on spatial queues arising from applicational demands in the field of mobile communication networks, it should not be forgotten that the use in other fields of application may turn out to be fruitful as well.

In spatial queueing theory, point fields can be used as a model for spatially distributed batch arrivals into a queue. Since arrivals in queues are always finite, we shall restrict ourselves to the theory of finite point fields. This section collects some classical results on stochastic point fields by Moyal [91] and Daley and Vere–Jones [48]. The first subsection describes the construction of point fields from a given set of finite–dimensional marginal distributions. Examples of point fields are given in the last subsection.

1.1 Definition and construction

The simplest way to see point fields is in terms of counting functions. A counting function yields for every subset A of the basic space R the number of elements of the point family that are contained in A.

Let $(R, \mathcal{R})$ be a measurable space with $\{x\} \in \mathcal{R}$ for every $x \in R$. A function $N : \mathcal{R} \to I\!N_0$ is called **counting function** if for any collection $(A_i : i \in I)$ of disjoint measurable subsets of R there are at most finitely many subsets $A_{i_1}, \dots, A_{i_n}$ for which $N(A_{i_k}) \geq 1$, and the equality

$$N\left(\sum_{i\in I} A_i\right) = \sum_{i\in I} N(A_i) = \sum_{k=1}^{n} N(A_{i_k})$$

holds. Let C_R denote the space of all counting functions on R.

Remark 7.1 If the number of points in the measurable set $A \in \mathcal{R}$ is represented by $N(A)$, then any finite family of points in R has a unique representation by a counting function N. This is only true because of the restriction that every singleton $\{x\}$ is contained in the σ-algebra $\mathcal{R}$ (see Moyal [91], p.9).

Define $\mathcal{C}$ as the smallest σ–algebra on C_R that contains all sets of the form

$$\{N \in C_R : N(A_i) = k_i \ \ \forall i \in \{1, \ldots, n\}\}$$

with $n \in I\!N$ and $A_i \in \mathcal{R}$ measurable, $k_i \in I\!N_0$ for all $i \in \{1, \ldots, n\}$. If Π is a probability measure on $(C_R, \mathcal{C})$, then $(C_R, \mathcal{C}, \Pi)$ shall be called **stochastic point field** (or shortly: point field).

In order to construct a point field on a space R from distributions on finitely many subsets of R, we will use results from Daley and Vere–Jones [48], chapter 6. Let R be a Polish space and $\mathcal{R}$ its Borel σ–algebra. Then every singleton $\{x\}$ is contained in $\mathcal{R}$. This follows from

$$\{x\} = \bigcap_{n=1}^{\infty} B\left(x, \frac{1}{n}\right)$$

with $B(x, d) := \{y \in R : d(x, y) < d\}$ denoting the set of points $y \in R$ having a distance less than d from x. That means that the condition on $\mathcal{R}$ appearing in the definition of counting functions is satisfied.

The first immediate result confirms the unique representation of a point field by a sufficiently large set of finite–dimensional marginal distributions. Define

$$\Pi^{S_1, \ldots, S_n}(k_1, \ldots, k_n) := \Pi(\{N \in C_R : N(S_i) = k_i \ \ \forall i \in \{1, \ldots, n\}\})$$

for all $n \in I\!N$, $S_1, \ldots, S_n \in \mathcal{R}$, and $k_1, \ldots, k_n \in I\!N_0$. According to Daley and Vere–Jones [48], proposition 6.2.III, the probability measure Π is uniquely determined by the marginal distributions $\Pi^{S_1, \ldots, S_n}$ on finite families of disjoint sets $S_1, \ldots, S_n \in \mathcal{R}$ from a semi-ring of bounded sets generating $\mathcal{R}$.

The main result states sufficient and necessary conditions for the existence of a point field, if the set of finite–dimensional marginal distributions is given. The first condition is merely notational. The second condition is the consistency condition for Kolmogorov's extension theorem, which ensures that a probability measure having the given marginal distributions can be constructed. The last two conditions guarantee that the support set of the constructed probability measure is exactly the space C_R of counting functions.

Theorem 7.2 *There exists a point field Π on $(R, \mathcal{R})$ with finite–dimensional distributions $\left(\Pi^{S_1, \ldots, S_n} : n \in I\!N, S_1, \ldots, S_n \in \mathcal{R}\right)$ if and only if the following conditions are satisfied:*

1 *For all* $n \in \mathbb{N}$ *and every permutation* $(i_1, \ldots, i_n)$ *of the integers* $(1, \ldots, n)$ *the equalities* $\Pi^{S_1,\ldots,S_n}(k_1, \ldots, k_n) = \Pi^{S_{i_1},\ldots,S_{i_n}}(k_{i_1}, \ldots, k_{i_n})$ *hold.*

2 *For every* $n \in \mathbb{N}$ *and* $S_1, \ldots, S_n \in \mathcal{R}$, *the consistency condition*

$$\Pi^{S_1,\ldots,S_{n-1}}(k_1, \ldots, k_{n-1}) = \sum_{k=0}^{\infty} \Pi^{S_1,\ldots,S_n}(k_1, \ldots, k_{n-1}, k)$$

holds.

3 *For every* $n \in \mathbb{N}$ *and* $S_1, \ldots, S_n \in \mathcal{R}$, *the measure* $\Pi^{S_1,\ldots,S_n}$ *has support* $\mathbb{N}_0^n$.

4 *For disjoint Borel sets* $A, B \in \mathcal{R}$ *and integers* $k, l \in \mathbb{N}_0$, *the equality*

$$\Pi^{A,B,A+B}(k, l, k+l) = \Pi^{A,B}(k, l)$$

holds.

Proof: The necessity is shown first: Condition 1 is a mere notational requirement. Condition 2 expresses the consistency requirement needed in Kolmogorov's extension theorem (see Gikhman and Skorokhod [62], p.108, theorem 3). If it were not satisfied, the given distributions could not be marginal distributions of the same measure. Condition 3 is necessary, because if for some measure $\Pi^{S_1,\ldots,S_n}$ the support would exceed $\mathbb{N}_0^n$, then the support of any measure on $(R, \mathcal{R})$ with marginal measure $\Pi^{S_1,\ldots,S_n}$ would exceed $\mathbb{N}_0^{\mathcal{R}}$ and hence the space $C_R \subset \mathbb{N}_0^{\mathcal{R}}$ of counting functions. Finally, if condition 4 fails to hold, any measure on $(R, \mathcal{R})$ with marginal measures $\Pi^{A,B,A+B}$ and $\Pi^{A,B}$ would have a support set that violates the second property in the definition of counting functions and hence is greater than the set of counting functions.

Now we show sufficiency: According to Kolmogorov's extension theorem, the first two conditions guarantee that a probability measure Π on $(R, \mathcal{R})$ with marginal measure $\Pi^{S_1,\ldots,S_n}$ can be constructed. Condition 3 implies that the support of Π is on $\mathbb{N}_0^{\mathcal{R}}$ and further

$$\sum_{n=0}^{\infty} \Pi(R)(n) = 1$$

which ensures the first property in the definition of counting functions. Condition 4 states that for any two disjoint Borel sets $A, B \in \mathcal{R}$,

$$\Pi(\{N \in C_R : N(A) + N(B) = N(A+B)\}) = 1$$

Since the Borel σ-algebra of a Polish space can be generated by a countable semi-ring (see Daley and Vere–Jones [48], p.608, theorem A2.1.III), this implies that the second property in the definition of counting functions holds almost surely with respect to Π. Hence, Π is a probability measure on the set

C_R of counting functions.
☺

An easy and constructive way to specify finite point fields can be found in Daley and Vere–Jones [48], chapter 5.3. By this method, first the probability p_n of the number n of points in R is given and then, conditioned on $n \geq 1$, a probability distribution Π_n of the locations of the points. The indistinguishability implies that every Π_n must be symmetric. Daley and Vere–Jones [48], p.126 (proposition 5.3.II), give the constructive

Theorem 7.3 *Let $(R, \mathcal{R})$ be a Polish space with its Borel σ-algebra. Then a finite point field is given by a discrete probability distribution $(p_n : n \in \mathbb{N}_0)$ and, for each integer $n \in \mathbb{N}$, a symmetric probability distribution Π_n on the σ-algebra $\mathcal{R}^n$ of R^n.*

1.2 Examples

Concluding this section, some examples of stochastic point fields are given. The first one is the well–known Poisson field, which assumes independence of the point positions. The next two examples are widely used variations of the Poisson field. For more examples see Cressie [47], Upton and Fingleton [123] or Diggle [51].

Example 7.4 Let $(R, \mathcal{R})$ be a measurable space with $\{x\} \in \mathcal{R}$ for every $x \in R$. If the distribution of points in disjoint sets is independent, i.e. if

$$\Pi^{S_1,\dots,S_n}(k_1, \dots, k_n) = \prod_{i=1}^{n} \Pi^{S_i}(k_i)$$

for any disjoint sets $S_1, \dots, S_n \in \mathcal{R}$, and if for all sets $S \in \mathcal{R}$ the distribution Π^S is a Poisson distribution with some parameter $\mu(S) \in [0, \infty]$, then the point field $(C_R, \mathcal{C}, \Pi)$ shall be called a **Poisson point field** on $(R, \mathcal{R})$. These are examined in Kingman [76]. Some important basic properties are given in the following theorems. First we obtain according to Kingman [76], p.11f,23, the fundamental

Theorem 7.5 *The function $\mu : \mathcal{R} \to [0, \infty]$, which yields the parameters $\mu(S)$ of the Poisson distributions Π^S, is a measure on $(R, \mathcal{R})$. On the other hand, if μ is a non–atomic measure on $(R, \mathcal{R})$ which can be expressed in the form $\mu = \sum_{n=1}^{\infty} \mu_n$ with $\mu_n(R) < \infty$ for every $n \in \mathbb{N}$, then there is a Poisson point field $(C_R, \mathcal{C}, \Pi)$ on $(R, \mathcal{R})$ with Π^S having parameter $\mu(S)$.*

Since $E(\Pi^S) = \mu(S)$, the measure μ is called the **mean measure** of $(C_R, \mathcal{C}, \Pi)$. An important example for a mean measure is the following: Let

ν be a non–atomic finite measure on $(R, \mathcal{R})$. Further, let $f : R \to I\!R_0^+$ be a bounded ν-measurable function. Then the measure μ defined by

$$\mu(A) = \int_A f(x) d\nu(x)$$

for all $A \in \mathcal{R}$ is a mean measure. The function f shall be called the **intensity function** with respect to ν. The value $f(x)$ is called **intensity** of $x \in R$.

Example 7.6 If the mean measure of a Poisson point field is allowed to be random in the same probability space as the point field itself, then the resulting point field shall be called a **Cox field**. For a more formal definition and properties of Cox fields, see Cressie [47], pp.657-661, or Grandell [64].

Example 7.7 Another point field which is derived from the Poisson point field is the **Poisson cluster field**. It is defined by the following construction:

1 Parent events are realized from a Poisson point field on R with some mean measure μ.

2 Each parent produces a random number of offspring.

3 The offspring are positioned arbitrarily in R, depending on the parent position.

4 The final process is composed of the superposition of offspring only.

A special kind of Poisson cluster fields is the **Neyman–Scott process.** A Poisson cluster field on $R = I\!R^d$ is called a Neyman–Scott process if the random number of offspring as well as the positions of the offspring relative to their parents are independently and identically distributed. For properties of Neyman–Scott processes, see Cressie [47], pp.662-669.

2. Definition and Construction of SMAPs

In this section, the spatial Markovian arrival process (or shortly SMAP) will be introduced. This shall be done by using the concept of stochastic point fields developed in the previous section. The space of counting functions on some arrival space will take the place of the additive space of a discrete Markovian arrival process. Thus, point fields serve as a description for the spatial distribution of arrivals for SMAPs and henceforth, they shall be called **arrival fields**.

After defining spatial Markovian arrival processes as a special case of discrete Markovian arrival processes, a way of constructing SMAPs from a given generator of the phase process and a set of finite–dimensional marginal distributions of the arrival field is described. The concept of marginal characteristic sequences as well as the parallel presentations in sections 3 and 4 should elucidate the familiarity with BMAPs.

Let R be a Polish space and $\mathcal{R}$ Borel's σ-algebra on R. Further, let C_R denote the space of counting functions on $(R, \mathcal{R})$. Let $\mathcal{C}$ be the σ-algebra on C_R as constructed in the previous section. Further, let Φ denote a separable metric space, which is locally compact, and let $\mathit{\Phi}$ denote a σ-algebra on Φ. A Markov–additive jump process (N, J) with state space $E := C_R \times \Phi$ shall be called **spatial Markovian arrival process** (or shortly **SMAP**) with **arrival space** $(R, \mathcal{R})$.

Remark 7.8 Since $C_R \subset I\!N_0^{\mathcal{R}}$, a spatial Markovian arrival process (N, J) is a special kind of discrete Markovian arrival processes. Hence we can call Φ the phase space and J the phase process of (N, J).

Remark 7.9 The batch Markovian arrival process (or shortly BMAP) developed by Lucantoni [80] is a homogeneous SMAP with trivial arrival space and a finite phase space $\Phi = \{1, \ldots, m\}$. Hence, all results for SMAPs and queues with SMAPs are valid for BMAPs and queues with BMAPs, too. Thus, chapter 5 actually is a special case of chapter 9.

The following results show a way of constructing an SMAP. First, conversely to theorem 2.5, a generator on the state space $C_R \times \Phi$ is composed of a kernel on Φ and a kernel from Φ to $\mathcal{C}$. After that, it is shown that the second kernel (which includes a measure on the space C_R of counting functions) can be determined uniquely by a sufficiently large set of finite–dimensional marginal distributions.

Theorem 7.10 *Let $t \to D_t^{\Phi}$ be a bounded and continuous generator function, determining a Markov jump process on Φ. Further, let $K_{t,y} : \Phi \times \mathcal{C} \to [0, 1]$ be Markov kernels such that the function $t \to K_{t,y}$ is continuous for every $y \in \Phi$. Define*

$$Q_t(y, A \times B) := \int_B K_{t,y}(z, A)\ D_t^{\Phi}(y, dz)$$

for all $t \in I\!R_0^+$, $y \in \Phi$, $A \in \mathcal{C}$ and $B \in \mathit{\Phi}$. Then $t \to Q_t$ is a bounded and continuous generator function that determines a spatial Markovian arrival process.

Proof: On the space of kernels on E, we have defined the supremum norm

$$||K|| := \sup_{x \in E, A \in \mathcal{E}} |K(x, A)|$$

for every kernel K on E. By assumption, the function $t \to D_t^{\Phi}$ is bounded. Because $K_{t,y}$ is a Markov kernel for every $y \in \Phi$ and $t \in I\!R_0^+$, this implies that the function $t \to Q_t$ is bounded, too.

Fix any $t \in I\!R_0^+$. Then the relation

$$
\begin{aligned}
&|Q_{t+h}(y, A \times B) - Q_t(y, A \times B)| \\
&\quad = \left| \int_B K_{t+h,y}(z, A)\ D^\Phi_{t+h}(y, dz) - \int_B K_{t,y}(z, A)\ D^\Phi_t(y, dz) \right| \\
&\quad \leq \left| \int_B K_{t+h,y}(z, A)\ D^\Phi_{t+h}(y, dz) - \int_B K_{t+h,y}(z, A)\ D^\Phi_t(y, dz) \right| \\
&\qquad + \left| \int_B (K_{t+h,y}(z, A) - K_{t,y}(z, A))\ D^\Phi_t(y, dz) \right| \\
&\quad \leq \|D^\Phi_{t+h} - D^\Phi_t\| + \|D^\Phi_t\| \cdot \|K_{t+h,y} - K_{t,y}\|
\end{aligned}
$$

implies the continuity of $t \to Q_t$, since $\|D^\Phi_t\|$ is bounded.
☺

In order to uniquely determine the measure part $K_{t,y}(z, .)$ of the kernel $K_{t,y}$, it suffices to specify the finite–dimensional marginal measures. For $n \in I\!N$, $S_1, \ldots, S_n \in \mathcal{R}$ and $k_1, \ldots, k_n \in I\!N_0$, define

$$pr^{-1}_{(S_1,\ldots,S_n)}(k_1, \ldots, k_n) := \{N \in C_R : N(S_i) = k_i\ \forall i \in \{1, \ldots, n\}\}$$

as the inverse image of the projection of functions in C_R to $(S_1, \ldots, S_n)$. Further, denote the stochastic point fields

$$\Pi_{t,y,z} := K_{t,y}(z, .)$$

and their finite–dimensional marginal distributions at $(S_1, \ldots, S_n)$ by

$$\Pi^{S_1,\ldots,S_n}_{t,y,z}(k_1, \ldots, k_n) := \Pi_{t,y,z}(pr^{-1}_{(S_1,\ldots,S_n)}(k_1, \ldots, k_n))$$

for all $t \in I\!R_0^+$ and $y, z \in \Phi$. From Daley and Vere–Jones [48], proposition 6.2.III, and theorem 7.10, we obtain immediately

Theorem 7.11 *An SMAP on $(R, \mathcal{R})$ is uniquely determined in distribution by the generator $t \to D^\Phi_t$ of its phase process and the marginal distributions $\Pi^{S_1,\ldots,S_n}_{t,y,z}$ of its arrival fields on finite families of disjoint subsets $S_1, \ldots, S_n \in \mathcal{R}$ from a semi-ring of bounded sets generating $\mathcal{R}$.*

The state space of the marginal process N is the space C_R of counting functions. Since a probability measure on C_R is uniquely determined by a sufficiently large set of finite–dimensional marginal distributions, it is possible to uniquely determine N by a set of finite–dimensional marginal processes of N. To this aim, define the stochastic processes

$$N^{S_1,\ldots,S_n} := pr_{(S_1,\ldots,S_n)}(N)$$

for all $n \in I\!N$ and $S_1, \ldots, S_n \in \mathcal{R}$. Then the process $(N^{S_1,\ldots,S_n}, J)$ shall be called the **marginal process** of (N, J) on $(S_1, \ldots, S_n)$.

In order to determine the transition probabilities of an SMAP (N, J), the transition probabilities of (N, J) will be given in terms of the transition probabilities of the marginal processes $(N^{S_1,\ldots,S_n}, J)$. Denote for any $n \in I\!N$ and $S_1, \ldots, S_n \in \mathcal{R}$ the marginal transition probabilities from time s to time $t > s$ by

$$P_{k_1,\ldots,k_n}^{S_1,\ldots,S_n}(s,t)(y,A) := P_{st}\left(y, pr^{-1}_{(S_1,\ldots,S_n)}(k_1,\ldots,k_n) \times A\right)$$

for every $k_1, \ldots, k_n \in I\!N_0$, $y \in \Phi$ and $A \in \mathit{\Phi}$. According to Daley and Vere–Jones [48], proposition 6.2.III, we obtain

Theorem 7.12 *The transition probabilities of an SMAP (N, J) are already determined by the transition probability kernels $P^{S_1,\ldots,S_n}(s,t)$ on finite families of disjoint subsets $S_1, \ldots, S_n \in \mathcal{R}$ from a semi-ring of bounded sets generating $\mathcal{R}$.*

For any $t \geq 0$, $n \in I\!N$, $S_1, \ldots, S_n \in \mathcal{R}$ and $k_1, \ldots, k_n \in I\!N_0$, define the kernels $D_{k_1,\ldots,k_n}^{S_1,\ldots,S_n}(t)$ on Φ by

$$D_{k_1,\ldots,k_n}^{S_1,\ldots,S_n}(t)(y,A) := Q_t\left(y, pr^{-1}_{(S_1,\ldots,S_n)}(k_1,\ldots,k_n) \times A\right) \tag{7.1}$$

for all $y \in \Phi$ and $A \in \mathit{\Phi}$. Then the sequence

$$\Delta^{S_1,\ldots,S_n}(t) := \left(D_{k_1,\ldots,k_n}^{S_1,\ldots,S_n}(t) : k_1, \ldots, k_n \in I\!N_0\right) \tag{7.2}$$

uniquely determines the marginal process $(N^{S_1,\ldots,S_n}, J)$. The time–dependent sequence $\Delta^{S_1,\ldots,S_n}(t)$ shall be called **characteristic sequence** of the marginal process $(N^{S_1,\ldots,S_n}, J)$. The same argument as above immediately yields

Theorem 7.13 *The distribution of an SMAP (N, J) is uniquely determined by the (time–dependent) characteristic sequences $\Delta^{S_1,\ldots,S_n}(t)$ of the marginal processes $(N^{S_1,\ldots,S_n}, J)$ on all finite families of disjoint subsets $S_1, \ldots, S_n \in \mathcal{R}$ from a semi–ring of bounded sets generating $\mathcal{R}$.*

Remark 7.14 The similarity between the form of the characteristic sequences of SMAPs and BMAPs is best seen by interpreting the sequence $\Delta(t)$ for BMAPs as the marginal characteristic sequence $\Delta^R(t)$ for an SMAP on the whole arrival space. This supports the intuitive notion of the difference between a BMAP and an SMAP: A BMAP cannot distinguish between different subsets of the arrival space.

In the following two sections, we shall derive expressions for the transition probability kernels $P^{S_1,\ldots,S_n}(s,t)$ in terms of the characteristic sequences

$\Delta^{S_1,\ldots,S_n}(t)$. First this will be shown for the simpler homogeneous case in the next section. In section 4, the general (i.e. possibly inhomogeneous) case will be examined.

3. The homogeneous case

Let (N, J) denote a homogeneous SMAP with arrival space $(R, \mathcal{R})$. Then for all $n \in \mathbb{N}$ and $S_1, \ldots, S_n \in \mathcal{R}$, we can define the characteristic sequences $\Delta^{S_1,\ldots,S_n} := \Delta^{S_1,\ldots,S_n}(t)$ which are constant in time $t \in \mathbb{R}_0^+$. Further, it suffices to find expressions for the transition probability kernels $P^{S_1,\ldots,S_n}(t) := P^{S_1,\ldots,S_n}(0,t)$ starting at the time origin.

Now fix $n \in \mathbb{N}$ and $S_1, \ldots, S_n \in \mathcal{R}$. In order to facilitate notations, we will from now on omit the superscript $S_1, \ldots, S_n$ whenever we need this space for other notations. To determine the transition probabilities of the marginal process $(N^{S_1,\ldots,S_n}, J)$, we first define

$$\Delta^{*0}_{k_1,\ldots,k_n} := \begin{cases} I, & \sum_{i=1}^n k_i = 0 \\ 0, & \text{otherwise} \end{cases}$$

with I denoting the identity kernel on Φ, and recursively

$$\Delta^{*m+1}_{k_1,\ldots,k_n} := \sum_{l_1=0}^{k_1} \cdots \sum_{l_n=0}^{k_n} \Delta^{*m}_{l_1,\ldots,l_n} D^{S_1,\ldots,S_n}_{k_1-l_1,\ldots,k_n-l_n}$$

for $m \in \mathbb{N}_0$. Now theorem 1.13, formula (2.5), and definition (7.1) imply that the probability of k_i arrivals in S_i (for $i = 1, \ldots, n$) during time t is given by

$$P^{S_1,\ldots,S_n}_{k_1,\ldots,k_n}(t) = \sum_{m=0}^{\infty} \frac{t^m}{m!} \Delta^{*m}_{k_1,\ldots,k_n} \tag{7.3}$$

using the above definition of the convolution operator $*$. For the proof it suffices to verify that

$$Q^m\left(y, pr^{-1}_{(S_1,\ldots,S_n)}(k_1,\ldots,k_n) \times A\right) = \Delta^{*m}_{k_1,\ldots,k_n}(y, A)$$

for all $m \in \mathbb{N}_0$, $y \in \Phi$, and $A \in \mathcal{\Phi}$, which is straightforward by induction on m. Thus representation (7.3) of the marginal transition probabilities for an SMAP is strikingly similar to the analogous expression (3.1) for homogeneous BMAPs. In fact, any one–dimensional marginal process (N^S, J) is a BMAP with general phase space, and any collection of one–dimensional marginal SMAPs is a collection of stochastically dependent BMAPs with general phase space.

As a useful corollary, which will be important for the estimation procedure in chapter 10, the definition of the convolution operator $*$ immediately yields

Theorem 7.15 *For a homogeneous SMAP with finite phase space, the distribution of the interarrival times on the finite family $(S_1, \ldots, S_n)$ of sets is phase–type and determined by*

$$P_{0,\ldots,0}^{S_1,\ldots,S_n}(t) = \exp\left(D_{0,\ldots,0}^{S_1,\ldots,S_n} \cdot t\right) := \sum_{m=0}^{\infty} \frac{t^m}{m!} \Delta_{0,\ldots,0}^{*m}$$

for all $t \geq 0$.

Thus if the distribution of the phases at time s is given by the probability vector π, then the time duration from s until the first arrival in $(S_1, \ldots, S_n)$ after s has phase–type distribution with representation $\left(\pi, D_{0,\ldots,0}^{S_1,\ldots,S_n}\right)$. For an introduction to phase–type distributions, see Neuts [96], chapter 2.

Concluding this section, we shall specify the results for expectations and asymptotic behaviour. Define the stochastic kernel

$$D := \sum_{k=0}^{\infty} D_k^R$$

According to equation (2.6) and definition (7.1), this is the generator of the phase process. Assume that a stationary probability measure π for D does exist, such that $\pi D = 0$ holds.

Now fix any set $S \in \mathcal{R}$ that we want to examine in more detail. Theorems 2.12 and 1.13 yield for the marginal expectation in S (which is a kernel on Φ here) the expression

$$E(N_t^S - N_s^S) = \int_s^t e^{D\cdot(u-s)} \sum_{k=1}^{\infty} k D_k^S \, e^{D\cdot(t-u)} \, du$$

for all $s < t$. If the process starts without prior arrivals and in phase equilibrium π, this becomes

$$E_\pi(N_t^S) = \int_\Phi d\pi(y) \sum_{n=1}^{\infty} k D_k^S(y, \Phi) \cdot t$$

for all $t > 0$. This simplification is due to the properties

$$\int_\Phi d\pi(y) e^{D\cdot(u-s)}(y, A) = \pi(A) \qquad \text{and} \qquad e^{D\cdot(t-u)}(y, \Phi) = 1 \qquad (7.4)$$

which hold for all $s < u < t$, $y \in \Phi$ and $A \in \mathcal{\Phi}$. Finally, theorem 2.16 yields the asymptotics

$$\frac{N_t^S}{t} \to \lambda(S) := \int_\Phi d\pi(y) \sum_{n=1}^{\infty} k D_k^S(y, \Phi) \qquad \text{as} \qquad t \to \infty$$

P–almost surely for all initial phase distributions. The term $\lambda(S)$ shall be called stationary arrival rate in S.

4. The general case

Now we shall derive the marginal transition probabilities in the general case. Remember from definition (7.1) that the kernels $D_{k_1,\ldots,k_n}^{S_1,\ldots,S_n}(t)$ contain the phase–dependent rates of observing k_i arrivals in the set S_i ($1 \leq i \leq n$) during the infinitesimal interval dt.

Fix any $n \in \mathbb{N}$ and $S_1, \ldots, S_n \in \mathcal{R}$. Again to facilitate notations, we will from now on omit the superscript $S_1, \ldots, S_n$ whenever we need this space for other notations. For all $s < t \in \mathbb{R}_0^+$ definition (2.3) is specified to

$$P_{k_1,\ldots,k_n}^{(0)}(s,t) := \begin{cases} I, & \sum_{m=1}^{n} k_m = 0 \\ 0, & \text{otherwise} \end{cases}$$

with I denoting the identity kernel on Φ. The recursion (2.4) translates to

$$P_{k_1,\ldots,k_n}^{(m)}(s,t) := \sum_{j_1=0}^{k_1} \cdots \sum_{j_n=0}^{k_n} \int_s^t P_{j_1,\ldots,j_n}^{(m-1)}(s,u) D_{k_1-j_1,\ldots,k_n-j_n}^{S_1,\ldots,S_n}(u)\ du$$

for all $k_1, \ldots, k_n \in \mathbb{N}_0$. With these definitions, the transition matrices are given by

$$P_{k_1,\ldots,k_n}^{S_1,\ldots,S_n}(s,t) = \sum_{m=0}^{\infty} P_{k_1,\ldots,k_n}^{(m)}(s,t)$$

for all $k_1, \ldots, k_n \in \mathbb{N}_0$. In order to obtain a closed form expression, we shall define the n–dimensional sequences

$$P^{S_1,\ldots,S_n}(s,t) := \left(P_{k_1,\ldots,k_n}^{S_1,\ldots,S_n}(s,t) : k_i \in \mathbb{N}_0\right)$$

for all $s < t$. Further, define a convolution C of two n–dimensional sequences A and B by its entries

$$C_{k_1,\ldots,k_n} := \sum_{j_1=0}^{k_1} \cdots \sum_{j_n=0}^{k_n} A_{j_1,\ldots,j_n} B_{k_1-j_1,\ldots,k_n-j_n}$$

for all $k_1, \ldots, k_n \in \mathbb{N}_0$. Then the transition matrices are given in closed form by

$$P_{k_1,\ldots,k_n}^{S_1,\ldots,S_n}(s,t) = \sum_{m=0}^{\infty} \underbrace{\int_0^t \int_0^{u_k} \cdots \int_0^{u_2}}_{\text{k integrals}} \Delta^{S_1,\ldots,S_n}(u_1) * \ldots * \Delta^{S_1,\ldots,S_n}(u_k)\ du_1 \ldots du_k \tag{7.5}$$

as follows by induction on the recursion depth m.

Analogously to remark 1.9, an iteration formula for the computation of the marginal transition kernels $P_{k_1,\dots,k_n}^{S_1,\dots,S_n}(s,t)$ is given by starting with

$$P_{k_1,\dots,k_n}^{[0]}(s,t) := \begin{cases} I, & \sum_{m=1}^{n} k_m = 0 \\ 0, & \text{otherwise} \end{cases}$$

and iterating by

$$P_{k_1,\dots,k_n}^{[m+1]}(s,t) := \sum_{j_1=0}^{k_1} \cdots \sum_{j_n=0}^{k_n} \int_s^t P_{j_1,\dots,j_n}^{[m]}(s,u) D_{k_1-j_1,\dots,k_n-j_n}^{S_1,\dots,S_n}(u)\, du + P_{k_1,\dots,k_n}^{[0]}(s,t)$$

for all $k_1, \dots, k_n \in \mathbb{N}_0$ and $m \in \mathbb{N}_0$. In the limit, this leads to the computation $P_{k_1,\dots,k_n}^{S_1,\dots,S_n}(s,t) = \lim_{m\to\infty} P_{k_1,\dots,k_n}^{[m]}(s,t)$.

As in the homogeneous case, the stochastic kernel

$$D(t) := \sum_{n=0}^{\infty} D_n^R(t)$$

is a generator for all $t \in \mathbb{R}_0^+$. The MJP with generator function $t \to D(t)$ shall be called the **phase process** of the SMAP. Denote the transition kernel of the phase process from time s to time $t \geq s$ by $P_{s,t}^{\Phi}$.

Again, we fix any set $S \in \Phi$ which might be of special interest. Then theorem 2.12 yields the expression

$$E(N_t^S - N_s^S) = \int_s^t P_{su}^{\Phi} \sum_{k=1}^{\infty} k D_k^S(u)\, P_{ut}^{\Phi}\, du \tag{7.6}$$

for the expectation kernel of the marginal process N^S in the set S over the time interval $]s,t]$. If the phase process has a stationary distribution π such that $\pi = \pi P_{0t}^{\Phi}$ for all $t \in \mathbb{R}^+$, then starting the phase process in this distribution without prior arrivals leads to expectation kernels

$$E_\pi(N_t^S) = \int_0^t \int_{\Phi} d\pi(y) \sum_{k=1}^{\infty} k D_k^S(u)(y,\Phi)\, du$$

for the number of arrivals in S until time $t \in \mathbb{R}^+$.

A **periodic SMAP** with period T shall be defined as an SMAP with the property $Q_{s+T} = Q_s$ for all $s \in [0,T[$. If the phase process of a periodic

SMAP has a periodic family $(\pi_t : t \in [0, T[)$ of asymptotic distributions, then theorem 2.14 yields for any set $S \in \mathit{\Phi}$

$$\frac{N_t^S}{t} \to \lambda(S) := \frac{1}{T}\int_0^T \int_{\Phi} d\pi_t(y) \sum_{k=1}^{\infty} kD_k^S(t)(y, \Phi)\, dt \qquad \text{as} \quad t \to \infty$$

P-almost surely for all initial phase distributions.

5. Examples

Throughout this section, all transition rates $D_{k_1,\ldots,k_n}^{S_1,\ldots,S_n}(y, A)$ which are not defined explicitly, shall be assumed as zero. For ease of notation only homogeneous processes are given. The extension to their inhomogeneous variants is straightforward by simply introducing a time–dependence of the parameters.

Example 7.16 If the phase space $\Phi = \{\varphi\}$ of an SMAP (N, J) is trivial, i.e. consisting of only one element, then (N, J) shall be called **Poisson process with spatial arrivals**. For these special SMAPs, the kernels $D_{k_1,\ldots,k_n}^{S_1,\ldots,S_n}$ on Φ become real numbers and all computations are much more tractable. In this case the representation

$$D_{k_1,\ldots,k_n}^{S_1,\ldots,S_n} = \lambda \cdot \Pi^{S_1,\ldots,S_n}(k_1, \ldots, k_n)$$

holds, with λ denoting the intensity of the Poisson process in time and Π denoting any stochastic point field. If Π equals a Poisson point field (see example 7.4), then the SMAP may be called **spatial Poisson process**. For these, the Poisson property is given in the combined spatio–temporal dimension, such that they are special Poisson point fields on the space $\mathbb{R}_0^+ \times R$.

Example 7.17 Let the arrival space R be separable and choose any upper bound $M \in \mathbb{R}^+$ for the arrival intensities at any point in R. Then the space $C(R)$ of continuous functions from R into the compact interval $[0, M]$ is a compact and separable metric space. This follows from standard topological results (cf. Herrlich [67], pp.206,118,117) and the fact that any continuous function is determined by its values on a countable dense subset of the separable space R. Define the phase space to be $\Phi := C(R)$ and let $\mathit{\Phi}$ denote the Borel σ-algebra on Φ. Since every function $f \in \Phi$ is continuous and hence uniquely determined by its values on a countable dense subset of R, it follows that $\{f\} \in \mathit{\Phi}$ for every $f \in \Phi$ (cf. Neveu [100], p.103). Phases out of this space can serve as intensity functions for arrivals distributed by inhomogeneous Poisson point fields (see example 7.4).

A **spatial Markov–modulated Poisson process** (or shortly **spatial MMPP**) can be defined as follows: Let D denote the generator of a homogeneous Markov jump process with state space Φ as described above and Π_f a Poisson point field on R (see example 7.4) with arrival intensity $\int_S f(x)d\nu(x)$ for $S \in \mathcal{R}$. Further, let $\gamma_f > 0$ for all $f \in \Phi$ such that $\sup_{f\in\Phi}\gamma_f < \infty$. The infinitesimal transition rates are defined as

$$D_{0,\dots,0}^{S_1,\dots,S_n}(f,A) := D(f,A) - \gamma_f\left(1 - \Pi_f^{S_1,\dots,S_n}(0,\dots,0)\right)$$

for $A \in \Phi$ and $f \in A$,

$$D_{0,\dots,0}^{S_1,\dots,S_n}(f,A) := D(f,A)$$

for $A \in \Phi$ and $f \notin A$ and

$$D_{k_1,\dots,k_n}^{S_1,\dots,S_n}(f,A) := \gamma_f \cdot \Pi_f^{S_1,\dots,S_n}(k_1,\dots,k_n)$$

for $\sum_{i=1}^n k_i \geq 1$, $A \in \Phi$ and $f \in A$.

Obviously, Π_f does not need to be a Poisson point field on R and could be chosen as any other modulated random point field instead. However, here we have chosen the Poisson field because we want a spatial Poisson process to satisfy the Poisson independence property (see Kingman [76]) as well in the temporal as in the spatial dimension.

Example 7.18 A general construction of homogeneous SMAPs with finitely many phases may proceed as follows: Let $\Delta' := (D'_n : n \in I\!N_0)$ denote the characteristic sequence of a homogeneous BMAP. For every $n \in I\!N_0$ and any combination $i, j \in \Phi$, define a symmetric distribution $\Pi_{n;ij}$ on R^n. This yields a stochastic point field which is supported by the set of counting functions C with $C(R) = n$, i.e. with exactly n arrivals in the arrival space R. Note that for $n = 0$ the distributions $\Pi_{0;ij}$ are trivial, namely the Dirac distribution on the empty point field. Now an SMAP is defined by

$$Q(i, A \times \{j\}) := \sum_{n=0}^{\infty} D'_{n;ij} \cdot \Pi_{n;ij}(A)$$

for all $i, j \in \Phi$ and $A \in \mathcal{C}$, in the terms of theorem 7.10. In the more constructive terms of finite–dimensional marginal distributions, this translates to the definitions

$$D_{k_1,\dots,k_n}^{S_1,\dots,S_n}(i,j) := \sum_{k=0}^{\infty} D'_{k;ij} \cdot \Pi_{k;ij}^{S_1,\dots,S_n}(k_1,\dots,k_n)$$

for all $i, j \in \Phi$, $n \in I\!N_0$, $k_i \in I\!N_0$ and $S_i \in \mathcal{R}$.

This constructive way of defining a class of SMAPs has been pursued in Baum and Kalashnikov [17]. They called the processes defined above **spatial BMAPs** (or shortly: SBMAPs). Because of theorem 7.3, this construction yields all homogeneous SMAPs with finitely many phases.

Chapter 8

THE $SMAP/M_T/C/C$ QUEUE

In most textbooks on queueing theory, Markovian queues are the first ones that are analyzed. This clearly is a result of the simple structure of Markovian queues, which can be analyzed as a Markov jump process in a straightforward manner. For the same reason, we shall first examine the Markovian spatial queues. However, this will be done on a more general level, covering the important case of periodic Markovian queues.

A typical property of communication traffic is the dependence of its arrival rates on time. This aspect incites the use of time–inhomogeneous processes and queues for modelling communication networks. Typically, a periodic dependence of the arrival rates and/or the service time distribution can be assumed with period lengths of a day or a week.

While queues with periodic input naturally reflect the time–dependent traffic that arrives in communication networks, the analysis of queues with inhomogeneous arrival rates is far less developed than the one for homogeneous queues. Some of the existing results in the literature are given in Asmussen and Thorisson [8], Bambos and Walrand [13], Falin [57], Harrison and Lemoine [65], Hasofer [66], Heyman and Whitt [68], Lemoine [79, 78], Rolski [112, 113], and Willie [128].

The queue examined in this chapter is defined as follows. The arrival process is an SMAP. The number c of servers is finite and equals the system capacity for users. This means that whenever an arriving user finds all servers busy, he cannot enter the system and is rejected. Hence there are no waiting users in the queue. Every server is equal, and the service time distribution function B_s for a user arriving at time $s \in \mathbb{R}_0^+$ is defined by $B_s(t) := 1 - e^{-\int_s^{s+t} \mu_u du}$ for all $t \geq 0$, with $\mu_u \in \mathbb{R}_0^+$ for all $u \in [s, t]$. Assume that the function $t \to \mu_t$ is continuous. This means that the service process without idle periods would be an inhomogeneous Poisson process with rates $(\mu_t : t \in \mathbb{R}_0^+)$.

Remark 8.1 The similar SMAP/PH/c/c queue with phase–type service time distribution can be analyzed using exactly the same methods as the ones shown in the following. However, notations would become more complicated because one needs to use Kronecker compositions, see chapter 4. This is the main reason for demonstrating the analysis by the simpler case.

The $SMAP/M_t/c/c$ queue is a natural model for a cell of FDMA or TDMA wireless communication networks. For such a cell, which remains of constant size over time (unlike a CDMA cell), the arrival space R of the queue would be the landscape covered by the cell. The number c of servers would be the number of available channels, i.e. the ratio of the total band width to a single frequency lot assigned to every user (in the FDMA case). The assumption that no user is waiting appears natural, since a telephone customer who finds a busy line will not wait but choose another net provider. Furthermore, usually the arrival process will be periodic with the period length being a day or a week.

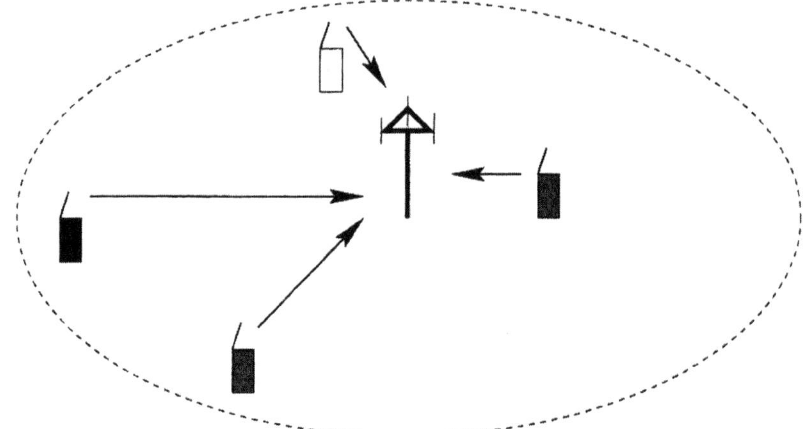

Model for a cell of a wireless communication network
(with four users and base station in the center)

In order to derive the simplest possible form of a loss formula (see section 3), the rejection policy of the communication network to be modelled shall be defined as follows. A rejection shall be conceived to occur only in the event that the system capacity is totally filled, i.e. that there are c users in the system. Thus, if there are $k < c$ users in the system and a batch arrival of size h with $k + h > c$ occurs, then this will not be rejected although it cannot be totally served by the system. In this case, the queue will be filled to its full capacity c. This can be interpreted as reflecting the interest of the network provider to partially serve an incoming user instead of rejecting him, even if the total amount of service that the user requires would exceed the system capacity. It is possible to derive transient and asymptotic distributions as well as a loss formula for any other rejection policy using the same methods, but this would result in more complicated formulae.

Let (N, J) denote the SMAP of the queue with arrival space $(R, \mathcal{R})$ and

$$\left(\Delta^{S_1,\dots,S_n}(t) : t \in I\!R_0^+, n \in I\!N, S_1, \dots, S_n \in \mathcal{R}\right)$$

denoting the marginal characteristic sequences of (N, J), see definition (7.2). Further, let $(\mu_t : t \in I\!R_0^+)$ denote the time–dependent service rates, which shall be equal for all servers. Denote the queue process by $Q = (Q_t : t \geq 0)$, with Q_t indicating the number of users in the system at time t.

Since Q is a spatial queue, any Q_t will denote a counting function on $(R, \mathcal{R})$. Analogous to theorem 7.12, from Daley and Vere–Jones [48], proposition 6.2.III, we obtain that the queue process is determined by its marginal processes $Q^{S_1,\dots,S_n}$ on all finite families of disjoint subsets $S_1, \dots, S_n \in \mathcal{R}$ from a semi–ring of bounded sets generating $\mathcal{R}$.

For ease of notation, the distribution of the queue process shall be derived only for one–dimensional marginal processes Q^S, i.e. only one subset $S \in \mathcal{R}$ will be described. For higher–dimensional marginal distributions on $(S_1, \dots, S_n)$ with $n > 1$, the method of analysis is completely analogous.

Denote the marginal queue process in the subsets R and S by $Q^{R,S} = \left(Q_t^{R,S} : t \in I\!R_0^+\right)$. Its infinitesimal generator $G_t^{R,S}$ can be written in a block matrix with entries being the kernels

$$G_t^{R,S}((h,k),(l,n)) = \begin{cases} k\mu_t \cdot I, & h = l+1 \leq c, \\ & k = n+1 \leq l+1 \\ (h-k)\mu_t \cdot I, & h = l+1 \leq c, \\ & k = n \leq l \\ D_{l-h,n-k}^{R,S}(t), & h < l < c, \\ & k \leq n \leq l \\ \sum_{i=c-h}^{\infty} D_{i,n-k}^{R,S}(t), & h < l = c, \\ & k \leq n \leq l \\ D_{0,0}^{R,S}(t) - h\mu_t \cdot I, & h = l < c, \\ & k = n \leq l \\ D_{0,0}^{R,S}(t) - h\mu_t \cdot I & h = l = c, \\ \quad + \sum_{i=1}^{\infty} D_i^R(t), & k = n \leq l \\ 0, & \text{otherwise} \end{cases}$$

for all $h, k, l, n \in \{0, \dots, c\}$, with 0 and I denoting the zero and identity kernel, respectively.

Here, the first two lines represent service completions within S or outside of S, respectively. The next two lines contain the rates for arrivals into the queue, the case in line 3 leaving some servers idle and the case in line 4 filling all system capacity. Finally, lines 5 and 6 represent the diagonal elements of the generator, representing phase changes without arrivals and service completions.

Remember that the $D_{l,n}^{R,S}(t)$ are kernels on the phase space Φ of the SMAP (N, J) with $D_{l,n}^{R,S}(t)(y, A)$ denoting the infinitesimal transition rate at time t in phase y to change to some phase $z \in A$ and observe n arrivals in S as well as l arrivals in R. Thus the above definition of the generator means that a phase transition may occur even if the arrival occuring at the same time is rejected.

1. Transition Probabilities

As a Markov jump process, the queue allows an explicit description of its transition probabilities in terms of the generator function. First, this shall be expressed by formula (1.6). Then, the obtained general form will be simplified for special cases.

The multiplication of two block matrixs K and L with kernel entries is defined by $KL((h,k),(l,n)) := \sum_{i=0}^{\infty}\sum_{j=0}^{\infty} K((h,k),(i,j))L((i,j),(l,n))$, which for the above generators can be simplified to

$$KL((h,k),(l,n)) = \sum_{i=0}^{c}\sum_{j=0}^{i} K((h,k),(i,j))L((i,j),(l,n))$$

for all $h, k, l, n \in \{0, \ldots, c\}$.

Define $P_{st}^{R,S}((h,k), y; (l,n), A)$ as the transition probability of having (l,n) users in (R,S) and being in some phase $z \in A$ at time $t > s$ under the condition of having (h,k) users in (R,S) and being in phase y at time s. Further define $P_{st}^{R,S}((h,k),(l,n))$ as the sub–stochastic kernel on the phase space Φ with entries $P_{st}^{R,S}((h,k), y; (l,n), A)$ and $P_{st}^{R,S}$ as the block matrix with kernel entries $P_{st}^{R,S}((h,k),(l,n))$.

Since the marginal queue process $Q^{R,S}$ is a Markov jump process, formula (1.6) applies to the transition probabilities and we obtain immediately

Theorem 8.2 *The transition probability kernel $P_{st}^{R,S}$ of the queue process $Q^{R,S}$ from time s to time t can be written as*

$$P_{st}^{R,S} = \sum_{k=0}^{\infty} \underbrace{\int_s^t \int_s^{u_k} \cdots \int_s^{u_2}}_{k\ integrals} G_{u_1}^{R,S} \cdots G_{u_k}^{R,S}\, du_1 \ldots du_k \tag{8.1}$$

defining the summand for $k = 0$ as the identity kernel on the state space $\mathbb{N}_0 \times \mathbb{N}_0 \times \Phi$ of $Q^{R,S}$.

The transition probabilities and an initial distribution yield immediately expressions for the transient distributions of the queue. Fix a time $t \in \mathbb{R}_0^+$ at which the queue is to be observed. Denote the initial distribution of the queue process at time $s := 0$ by ν. Then the transient distribution is determined by

$$P_t^{R,S} = \int d\nu P_{0t}^{R,S}$$

at all times $t > 0$.

Expression (8.1) assumes a simpler form for the following special cases. The simplest form can be arrived at in the homogeneous case. Then the transition probabilities assume the usual exponential form which holds for homogeneous Markov jump processes. Thus an immediate corollary from theorem 1.13 is

Theorem 8.3 *For homogeneous arrival and service rates, the generator $G := G^{R,S} := G_t^{R,S}$ is constant in t and the transition kernels from time s to time $t > s$ is determined by*

$$P_{st}^{R,S} = \sum_{k=0}^{\infty} \frac{(t-s)^k}{k!} G^k =: e^{G \cdot (t-s)}$$

with G^k denoting the k-th power of the matrix $G^{R,S}$.

Another case of simplification is that of quasi-commutable generators (see definition (1.15)). According to theorem 1.17, an exponential form is assumed, too, although with a more complicated exponent:

Theorem 8.4 *If the generators $(G_t^{R,S} : t \in \mathbb{R}_0^+)$ are quasi–commutable, which means that the equation*

$$\frac{d}{dt}\left(\int_0^t G_u^{R,S} du\right)^k = k \cdot \left(\int_0^t G_u^{R,S} du\right)^{k-1} G_t^{R,S}$$

holds for all $k \in \mathbb{N}$ and $t \in \mathbb{R}_0^+$, the transition probabilities of $Q^{R,S}$ are given by

$$P_{st}^{R,S} = \sum_{k=0}^{\infty} \frac{\left(\int_s^t G_u^{R,S} du\right)^k}{k!} = e^{\int_0^t G_u^{R,S} du}$$

A simple form of quasi–commutability is given if the generators have the form

$$G_t^{R,S} = \lambda(t) \cdot G^{R,S}$$

for all $t \in \mathbb{R}_0^+$, with $\lambda : \mathbb{R}_0^+ \to \mathbb{R}_0^+$ being a non–negative function in time and $G^{R,S}$ denoting a generator which is constant in time (cf. remark 1.16). This would for example be the case if the arrival process satisfied $\Delta(t) = \lambda(t) \cdot \Delta$, with Δ being constant in time, and the service rates were constant.

Most important for applications in modelling mobile communication networks is the case of periodic arrival and service rates. In this case, a computation of the transient distribution can be simplified as follows. The periodicity of the generator yields

$$P_{0,nT} = P_{0,(n-1)T}P_{(n-1)T,nT} = P_{0,(n-1)T}P_{0,T} = P_{0,T}^n$$

for all $n \in \mathbb{N}$. Define

$$\lfloor t/T \rfloor := \max\{n \in \mathbb{N}_0 : nT \leq t\}$$

as the number of period lengths that have passed until time $t \in \mathbb{R}^+$. Now the transient distributions of Q are determined by

$$P_{0t} = P_{0,\lfloor t/T \rfloor T}P_{\lfloor t/T \rfloor T,t} = (P_{0T})^{\lfloor t/T \rfloor} P_{0,t-\lfloor t/T \rfloor T}$$

for all $t > 0$. This expression allows a computation of the transient distribution at any time $t \in \mathbb{R}^+$ without needing to integrate over ranges larger than the period T. For computing the remaining terms P_{0s} with $s \leq T$, one can use iteration (1.7).

Example 8.5 Assume that the parameters are given as follows: Let Δ denote the characteristic sequence of a homogeneous SMAP and choose service rates $\mu_1, \mu_2 \in \mathbb{R}^+$. Further, let $T > 0$ denote a period length and set

$$\Delta(t) = \left(1 + \sin\left(\frac{2\pi}{T}t\right)\right) \cdot \Delta$$

as well as

$$\mu_t = \begin{cases} \mu_1 & \text{for} \quad t < T/2 \\ \mu_2 & \text{for} \quad t \geq T/2 \end{cases}$$

for all $t \in [0, T[$. Then the generators are periodic with period T and piecewise quasi–commutable (on $[0, T/2[$ and on $[T/2, T[$). Denote $G_1 := G_0^{R,S}$ as well as $G_2 := G_{T/2}^{R,S}$. Acccording to example 1.20, we then obtain

$$P_{0s} = \exp\left(\left(s + \frac{T}{2\pi}\left(1 - \cos\left(\frac{2\pi}{T}s\right)\right)\right) \cdot G_1\right)$$

for $s < T/2$ and

$$P_{0s} = \exp\left(\left(\frac{T}{2} + \frac{T}{\pi}\right) G_1\right)$$
$$\times \exp\left(\left(s - \frac{T}{2} - \frac{T}{2\pi}\left(1 + \cos\left(\frac{2\pi}{T}s\right)\right)\right) \cdot G_2\right)$$

for $s \geq T/2$.

2. Asymptotic Distributions

In the case of a periodic $SMAP/M_t/c/c$ queue with finite phase space, a periodic family of asymptotic distributions can be given. Let Q denote a periodic $SMAP/M_t/c/c$ queue with finite phase space $\Phi = \{1, \ldots, m\}$. Further, let $S \in \mathcal{R}$ denote any subset of the arrival space and $Q^{R,S}$ the marginal process of Q on the subsets R and S. Assume that $Q^{R,S}$ is periodic with period length $T > 0$.

Let $Y^{R,S} = (Y_n^{R,S} : n \in I\!N_0)$ be the homogeneous discrete time Markov chain with transition kernel

$$P_{0T}^{R,S} = \sum_{n=0}^{\infty} \int_0^T \int_0^{u_n} \ldots \int_0^{u_2} G_{u_1}^{R,S} \ldots G_{u_n}^{R,S} du_1 \ldots du_n$$

Then the main result is

Theorem 8.6 *The chain $Y^{R,S}$ has an asymptotic distribution $Q^{R,S}$, and the marginal queue process $Q^{R,S}$ has a periodic family $(q_s^{R,S} : s \in [0, T[)$ of asymptotic distributions (see definition 1.21) assuming the form*

$$q_s^{R,S} = q^{R,S} P_{0s}^{R,S}$$

for all $s \in [0, T[$. The distribution $q^{R,S}$ is determined by the equilibrium equation

$$q^{R,S} = q^{R,S} P_{0T}^{R,S}$$

Proof: Since the phase space Φ is finite and the queue has finite capacity, the state space of the queue process is finite, too. It follows that the Markov chain $Y^{R,S}$ has finite state space and hence an asymptotic distribution $q^{R,S}$. The rest follows from theorem 1.23.
☺

The homogeneous $SMAP/M/c/c$ queue can be seen as a special case of the periodic $SMAP/M_t/c/c$ queue with arbitrary period length $T > 0$ and generators $G^{R,S}$ which are constant in time. This leads to

Theorem 8.7 *Let Q be a homogeneous SMAP/M/c/c queue with finite phase space $\Phi = \{1, \ldots, m\}$ and generator $G := G^{R,S}$ for the marginal process on the sets R and S. Then $Q^{R,S}$ has an asymptotic distribution $q^{R,S}$ with*

$$q^{R,S} G^{R,S} = 0$$

Proof: This follows from the above theorem 8.6 and the form

$$P_{0t}^{R,S} = e^{G^{R,S} \cdot t} = \sum_{n=0}^{\infty} \frac{t^n}{n!} G^n$$

for the transition probability kernel of $Q^{R,S}$ (see theorem 1.13). The homogeneous Markov chain $Y^{R,S}$ defined above has an asymptotic distribution $Q^{R,S}$, since the state space is finite. This satisfies $q^{R,S} G^{R,S} = 0$ and because of the above exponential form $q^{R,S} P_{0s}^{R,S} = q^{R,S}$ for all $s \in [0, T[$.
☺

Remark 8.8 In the homogeneous case, the generator matrix

$$G^{R,S} = \left(G^{R,S}((h,k),(l,n)) \right)_{0 \leq k \leq h \leq c, 0 \leq n \leq l \leq c}$$

can be arranged according to the indices (k, n). This results in an $M/G/1$–type matrix and hence the distribution $q^{R,S}$ can be found by the method introduced in Ramaswami [108] and generalized by Hofmann [69, 70].

3. A Loss Formula

The $SMAP/M_t/c/c$ queue is a **loss system**. This means that arriving users who find all servers busy are rejected and can be regarded as lost. One of the most important performance measures of loss systems is the loss probability, which is defined as the long time fraction of lost, i.e. rejected users to all users having arrived. For the application field of mobile communcation networks, the meaning of this probability as a criterion of quality is immediate.

For $M/G/c/c$ loss systems, the loss probability is the asymptotic probability p_c of having c users in the system, which is the same event as all servers being busy. This result is independent of the service time distribution and follows easily from the PASTA property for queues with Poisson arrival processes (cf. Wolff [129], pp.271-273). For more recent results on non–spatial loss systems, see also the work of Willie [125, 126, 127, 128] who was able to obtain results for periodic classical queues. In this subsection, the loss probability will be derived using the law of large numbers for periodic SMAPs (theorem 2.14). It turns out that an analogous result holds, although the PASTA property does not hold for the $SMAP/M_t/c/c$ queue (for a more general treatment of the PASTA problem see Melamed and Yao [87]).

Let Q denote a loss system. Denote the number of users, that have arrived at Q until time $t \in I\!R_0^+$, by N_t and the number of rejected users until time $t \in I\!R_0^+$ by L_t. Assume that Q has an asymptotic distribution. Then the **loss probability** p_l of Q is defined as the fraction

$$p_l := \lim_{t\to\infty} \frac{L_t}{N_t}$$

of the number of rejected users over the number of all users having arrived in the long run.

Now let Q denote the periodic $SMAP/M_t/c/c$ queue with arrival space $(R, \mathcal{R})$, finite phase space $\Phi = \{1, \ldots, m\}$ and period length $T > 0$. As usual, notational convenience is the reason for analyzing only the one–dimensional marginal processes on any measurable set $S \in \mathcal{R}$. Denote the SMAP arrival process on the subset S by (N^S, J) and define it by its characteristic sequence $(D_n^S(t) : l, n \in I\!N_0)$. Then $D(t) := \sum_{n=0}^{\infty} D_n^R(t)$ is the generator of the phase process J at time t. This has a periodic family $(\pi_s : s \in [0, T[)$ of asymptotic distributions, since the phase space Φ is finite. Furthermore, the finiteness of Φ allows to represent the measures π_s as row vectors in $I\!R^m$ and the kernels $D_n^S(t)$ and $D(t)$ as $m \times m$ matrices with real-valued entries. Define the column vector $1_m \in I\!R^m$ as the one with all entries being 1. Finally, assume that the asymptotic mean arrival rate during one period length

$$\lambda(S) := \frac{1}{T}\int_0^T \pi_t \sum_{n=1}^{\infty} n D_n^S(t) 1_m dt < \infty$$

is finite. Then the main result is

Theorem 8.9 *Let Q denote a periodic $SMAP/M_t/c/c$ loss system with finite phase space $\Phi = \{1, \ldots, m\}$, arrival space $(R, \mathcal{R})$ and period length $T > 0$. Denote the characteristic sequence of the SMAP on the subset $S \in \mathcal{R}$ by $(D_n^S(t) : n \in I\!N_0)$. Assume that Q^R has a periodic family $(q_s : s \in [0, T[)$ of asymptotic distributions and define $q_t(c) \in I\!R^m$ as the row vector with ith entry $q_t(c, i)$. Then the loss probability of Q on the subset S is*

$$p_l(S) = \frac{\int_0^T q_t(c) \sum_{n=1}^{\infty} n D_n^S(t) 1_m dt}{\int_0^T \pi_t \sum_{n=1}^{\infty} n D_n^S(t) 1_m dt}$$

for all $S \in \mathcal{R}$.

Proof: For ease of notation, we shall write $N_t := N_t^S$, $L_t := L_t^S$ and $D_n(t) := D_n^S(t)$ in this proof. By definition, the arrival process N^S in S is a one–dimensional Markovian arrival process. Thus, the strong law of large numbers (theorem 2.14) applies and yields

$$\lim_{t\to\infty}\frac{N_t}{t}=\frac{1}{T}\int_0^T \pi_t \sum_{n=1}^{\infty} nD_n(t)1_m dt \tag{8.2}$$

almost surely.

The process $L^S=(L_t : t\in \mathbb{R}_0^+)$ which counts the number of rejected users can be described by a thinned Markovian arrival process. The infinitesimal transition rates of L^S are given by

$$\begin{aligned}
R_n(t;i,j) &= \lim_{\Delta t\to 0}\frac{P(Q_t^R=(c,i), N_{t+\Delta t}-N_t=n, J_{t+\Delta t}=j|J_t=i)}{\Delta t}\\
&= \lim_{\Delta t\to 0}\frac{P(N_{t+\Delta t}-N_t=n, J_{t+\Delta t}=j|Q_t^R=(c,i), J_t=i)}{\Delta t}\\
&\quad \cdot P(Q_t^R=(c,i)|J_t=i)\\
&= \lim_{\Delta t\to 0}\frac{P(N_{t+\Delta t}-N_t=n, J_{t+\Delta t}=j|J_t=i)}{\Delta t}\\
&\quad \cdot P(Q_t^R=(c,i)|J_t=i)\\
&= D_n(t;i,j)\cdot P(Q_t^R=(c,i)|J_t=i)
\end{aligned}$$

for every $n\in \mathbb{N}$, since the arrival process (N^S,J) only depends on the phase but not on the number of users in R. Furthermore, we have

$$R_0(t;i,j)=D_0(t;i,j)+\sum_{n=1}^{\infty} D_n(t;i,j)\cdot(1-P(Q_t^R=(c,i)|J_t=i))$$

Writing $t=kT+s$ and letting $k\to\infty$, the asymptotic rates are

$$\lim_{k\to\infty} R_n(kT+s;i,j)=D_n(s;i,j)\cdot\frac{q_s(c,i)}{\pi_s(i)}$$

for all $s\in[0,T[$, $n\in\mathbb{N}$ and $i,j\in\Phi$ and

$$\lim_{k\to\infty} R_0(kT+s;i,j)=D_0(s;i,j)+\sum_{n=1}^{\infty} D_n(s;i,j)\cdot\left(1-\frac{q_s(c,i)}{\pi_s(i)}\right)$$

for all $s\in[0,T[$ and $i,j\in\Phi$. Obviously, the generator of the phase process of L^S equals

$$\sum_{n=0}^{\infty} R_n(t;i,j)=D(t;i,j)$$

which is the generator of J. Hence, by theorem 2.14, the strong law of large numbers assures the convergence

$$\lim_{t\to\infty} \frac{L_t}{t} = \frac{1}{T}\int_0^T \pi_t \sum_{n=1}^{\infty} nR_n(t)1_m dt = \frac{1}{T}\int_0^T q_t(c) \sum_{n=1}^{\infty} nD_n(t)1_m dt \tag{8.3}$$

almost surely.

Combining the results 8.2 and 8.3, the loss probability is determined by

$$p_l(S) = \lim_{t\to\infty} \frac{L_t}{t}\cdot\frac{t}{N_t} = \frac{\int_0^T q_t(c)\sum_{n=1}^{\infty} nD_n(t)1_m dt}{\int_0^T \pi_t \sum_{n=1}^{\infty} nD_n(t)1_m dt}$$

for all $S \in \mathcal{R}$.

☺

As usual, the case of a homogeneous queue can be regarded as the special periodic queue with arbitrary period length $T > 0$ and a generator which is constant in time. Thus, the result for homogeneous queues follows immediately from the above theorem:

Theorem 8.10 *Let Q denote a homogeneous SMAP/M/c/c loss system with arrival space $(R, \mathcal{R})$. Let the SMAP of Q have a characteristic sequence $(D_n^S : n \in \mathbb{N}_0)$ on the subset $S \in \mathcal{R}$. Assume that Q^R has an asymptotic distribution q and define $q(c) \in \mathbb{R}^m$ as the row vector with ith entry $q(c, i)$. Then the loss probability of Q on the subset S is*

$$p_l(S) = \frac{q(c)\sum_{n=1}^{\infty} nD_n^S 1_m}{\pi \sum_{n=1}^{\infty} nD_n^S 1_m}$$

for all $S \in \mathcal{R}$.

Remark 8.11 The classical Erlang loss formula can be derived from the above formula via the following specifications: Since Erlang's queueing models were non–spatial, we obtain the formula

$$p_l = \frac{q(c)\sum_{n=1}^{\infty} nD_n 1_m}{\pi \sum_{n=1}^{\infty} nD_n 1_m}$$

which is an Erlang loss formula for homogeneous queues with BMAP input. Reducing the number of phases to one (i.e. $m := 1$), we obtain $\pi = 1$, and the expressions $\sum_{n=1}^{\infty} nD_n$ and $q(c)$ are positive numbers. Thus we obtain

$$p_l = \frac{q(c)\sum_{n=1}^{\infty} nD_n}{\sum_{n=1}^{\infty} nD_n} = q(c)$$

which is the classical Erlang loss formula, valid also for batch arrivals.

Chapter 9

SPATIAL QUEUES WITH INFINITELY MANY SERVERS

Spatial Queues with infinitely many servers arise naturally as models for the planning process of mobile communication networks. A very useful concept has been developed by Çinlar [44], partly on the basis of Massey and Whitt [84]. This assumes single arrivals distributed in time as a non–homogeneous Poisson process. Every arrival chooses a position in space according to a (time–dependent) distribution on the space and independent of all other users. Since the arrival process is a multi–dimensional Poisson process on the product space of time and arrival space, the queue process is Poisson as well, and the mean measures can be computed in a straightforward manner. This yields a nice solution to the queue process.

Mobile communication networks transmitting parallel data streams of different types (e.g. voice, data, video) and group arrivals which are not independent in space require a more general treatment. As the BMAP concept (see chapter 3) shows for non–spatial queueing models, parallel data types can be modeled by using phases. This will be done here, too. Furthermore, phases can be used to model special system characteristics. Spatial dependence in the positioning of group arrivals can be acknowledged by using Spatial Markovian Arrival Processes (SMAPs) as introduced in chapter 7.

In this chapter, spatial infinite server queues shall be examined in three stages of generality. Essentially, the same method of determining the distribution of the queue process applies to all of them. In a first step (section 1), this method will be shown for the simplest model of time–homogeneous arrival rates and non–moving users. A first generalization in section 2 treats general (i.e. possibly non–homogeneous) arrival rates, but still non–moving users. In section 3, a last step provides for user movement as well general arrival rates. Finally, section 4 contains an application of the results to the planning procedure for the positioning of base stations in a mobile communication network.

1. Homogeneous Arrival Rates without User Movements

In this section, spatial queues with homogeneous arrival rates and constant user positions are examined. Let Q denote a queue with an SMAP arrival process (N, J) and infinitely many independent servers. Denote the arrival space by $(R, \mathcal{R})$ and let $\{\Delta^{S_1,\ldots,S_n} : n \in I\!N, S_i \in \mathcal{R}\}$ be the marginal characteristic sequences determining (N, J). Further, let G denote the service time distribution function. Every incoming user is served immediately, i.e. there is no waiting time and the queue length is always zero.

For an introduction of the method of analysis, the service times first will be assumed to be general and iid, i.e. they do not depend on the user's position. At the end of this section it will be shown how to extend the model as to include service time distributions depending on the user's position.

1.1 One–dimensional marginal distributions

For ease of notation, the queue will be observed in only one subset $S \in \mathcal{R}$ first. Observations in finitely many subsets will be derived in section 1.2. Define $P_{st}^S(i)(y, A)$ as the probability of observing i users being served in S at time t and the system being in some phase $z \in A$ under the condition that at time $s < t$ there were no users being served in S and the system was in phase $y \in \Phi$. Let $P_{st}^S(i)$ denote the kernel on Φ with entries $P_{st}^S(i)(y, A)$. Further define the sequence $P_{st}^S := (P_{st}^S(i) : i \in I\!N_0)$ and set $P_t^S := P_{0t}^S$.

Let $t \in I\!R^+$ denote the time instant and $S \in \mathcal{R}$ the subset the queue is observed at. In any infinitesimal time interval $du = \lim_{h\to 0} \,]u, u+h]$ with $u \in [0, t[$, batch arrivals occur with rates $\Delta^S = (D_i^S : i \in I\!N_0)$ according to the SMAP arrival process. Given that an arrival of batch size $m \in I\!N$ occured during $]u, u+h]$ with $u \in [0, t[$, the probability of $i \in \{0, \ldots, m\}$ arrivals still being served at time t is distributed binomially by

$$\binom{m}{i} G^c((t-u)-)^i G((t-u)-)^{m-i},$$

denoting $G((t-u)-) := \lim_{h\to 0} G(t-(u+h))$ and $G^c(t) := 1 - G(t)$.

Conditioning upon the batch arrival rates, the total rate of i arrivals during a time interval du which are still in service at time t is

$$R_i^S(u,t) = \sum_{m=i}^{\infty} D_m^S \binom{m}{i} G^c((t-u)-)^i G((t-u)-)^{m-i} \tag{9.1}$$

for every $u \in [0, t[$ and $i \in I\!N_0$.

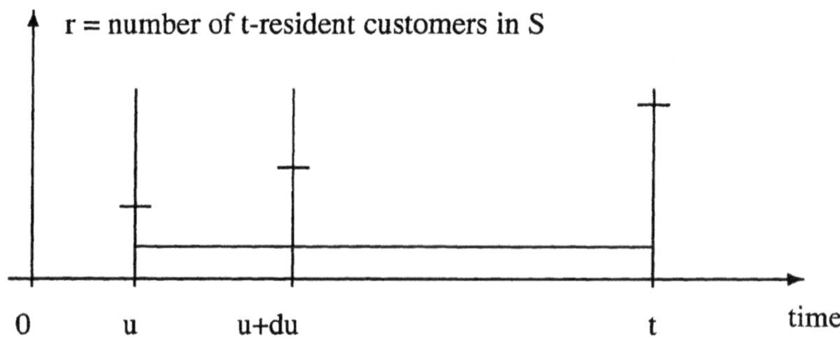

The sequence $R^S(u,t) := (R_i^S(u,t) : i \in I\!N_0)$ can be interpreted as the time–dependent marginal characteristic sequence of an inhomogeneous SMAP $H^{(t;S)} = (H_u^{(t;S)} : u \leq t)$ which at time $u = t$ coincides with the infinite server queue that is to be examined.

Note that for every time $t \in I\!R_0^+$ the queue is observed at, the sequence $R^S(u,t)$ and hence the SMAP $H^{(t;S)}$ is different. The transient distribution of the marginal queue process in the subset $S \in \mathcal{R}$ at time $t \in I\!R_0^+$ is determined by $P_t^S = H_t^{(t;S)}$. Formula (7.5) yields the following representation:

Theorem 9.1 *The transient distribution of the marginal queue process of Q in the subset $S \in \mathcal{R}$ at time $t \in I\!R_0^+$ is determined by*

$$P_t^S = \sum_{k=0}^{\infty} \underbrace{\int_0^t \int_0^{u_k} \cdots \int_0^{u_2}}_{k \text{ integrals}} R^S(u_1,t) * \ldots * R^S(u_k,t)\, du_1 \ldots du_k \tag{9.2}$$

defining the case of zero integrals as the sequence $Id = (I, 0, 0, \ldots)$ with I denoting the identity kernel and 0 the zero kernel on Φ, and further defining the convolution sequence $C = A * B$ by

$$C_n := \sum_{i=0}^{n} A_i B_{n-i}$$

for any two sequences A, B of kernels on Φ.

A great disadvantage in the above formula (9.2) lies in the fact that the sequences $R^S(u,t)$ need to be determined separately for every time $t \in I\!R_0^+$ the queue shall be observed at. In the next theorem, a closed form for all times of observance is obtained:

Theorem 9.2 *The transient distribution of the marginal queue process of Q in the subset $S \in \mathcal{R}$ at time $t \in \mathbb{R}_0^+$ is determined by*

$$P_t^S = \sum_{k=0}^{\infty} \underbrace{\int_0^t \int_0^{u_1} \dots \int_0^{u_{k-1}}}_{k \text{ integrals}} \tilde{R}^S(u_1) * \dots * \tilde{R}^S(u_k)\, du_k \dots du_1 \qquad (9.3)$$

with

$$\tilde{R}^S(u) := R^S(t-u, t) \qquad (9.4)$$

being independent of $t > u$.

Proof: By equation (9.1), the rates

$$\tilde{R}_i^S(u) := R_i^S(t-u,t) = \sum_{m=i}^{\infty} D_m^S \binom{m}{i} G^c(u-)^i G(u-)^{m-i}$$

are independent of $t > u$. Starting from equation (9.2), we obtain

$$\begin{aligned} P_t^S &= \sum_{k=0}^{\infty} \underbrace{\int_0^t \int_0^{u_k} \dots \int_0^{u_2}}_{\text{k integrals}} R^S(u_1,t) * \dots * R^S(u_k,t)\, du_1 \dots du_k \\ &= \sum_{k=0}^{\infty} \int_0^t \int_{u_1}^t \dots \int_{u_{k-1}}^t R^S(u_1,t) * \dots * R^S(u_k,t)\, du_k \dots du_1 \\ &= \sum_{k=0}^{\infty} \int_0^t \int_0^{u_1} \dots \int_0^{u_{k-1}} R^S(t-u_1,t) * \dots * R^S(t-u_k,t) du_k \dots du_1 \end{aligned}$$

which proves the statement.
☺

Remark 9.3 The representation (9.3) is from the point of view that one looks backward in time from time t until time 0. Thus, equation (9.3) gives the transition kernel of a process running backward in time with the phase process still running forward.

Example 9.4 Assume that the service time distribution is deterministic, i.e. $G = \delta_s$ is the Dirac measure with support $s > 0$. Then

$$\tilde{R}_i^S(u) = \begin{cases} D_i^S & \text{for} \quad u < s \\ \delta_{i,0} \cdot D & \text{for} \quad u \geq s \end{cases}$$

for all $i \in I\!N_0$, denoting the Kronecker function by δ and the generator of the phase process by $D = \sum_{n=0}^{\infty} D_n^S$. Thus we obtain $\tilde{R}^S(u) = \Delta^S$ for $u < s$ and $\tilde{R}^S(u) = D \cdot Id = (D, 0, 0, \ldots)$ for $u \geq s$ and hence

$$P_t^S = \begin{cases} e^{\Delta^S \cdot t} & \text{for} \quad t < s \\ P^{\Phi}_{0,t-s} e^{\Delta^S \cdot s} & \text{for} \quad t \geq s \end{cases}$$

for all $t \in I\!R_0^+$, with P^Φ denoting the transition probability kernel of the phase process.

The state space of the marginal process Q^S is $I\!N_0 \times \Phi$, hence the same as that of the marginal SMAP (N^S, J). Further, the phase process of the queue coincides with that of the SMAP, since the servers do not influence it. Hence we can define an expectation kernel $E(Q_t^S - Q_s^S)$ of the marginal queue process Q^S over the time interval $]s, t]$ exactly as we have defined one for MAJPs, namely by

$$E(Q_t^S - Q_s^S)(y, A) := E\left((N_t - N_s) \cdot 1_B(J_t) | J_s = y\right)$$

for all $s < t$, $y \in \Phi$ and $A \in \mathit{\Phi}$. Naturally, we write $E(Q_t^S) := E(Q_t^S - Q_0^S)$ with the understanding that $Q_0^S = 0$.

The next two theorems yield expressions for the expectation kernel of the marginal queue process Q^S at any time of observance:

Theorem 9.5 *Assume that the arrival rate in S is finite for any phase $y \in \Phi$, i.e.*

$$\sum_{n=1}^{\infty} n D_n^S(y, \Phi) < M < \infty \tag{9.5}$$

for all $y \in \Phi$. Then the expectation kernel of the marginal queue process Q^S in the subset $S \in \mathcal{R}$ at time $t \in I\!R_0^+$ is given by

$$E(Q_t^S) = \int_0^t P^{\Phi}_{0u} \sum_{n=1}^{\infty} n D_n^S P^{\Phi}_{ut} \, G^c((t-u)-) \, du \tag{9.6}$$

with P^Φ denoting the transition kernel of the phase process.

Proof: Fix a time $t \in I\!R_0^+$. Since P_t^S equals the distribution of the SMAP $H^{(t;S)}$ at time t, the expectation kernel at time t is given by equation (7.6) as

$$\begin{aligned} E(Q_t^S) &= \int_0^t P^{\Phi}_{0u} \sum_{n=1}^{\infty} n R_n^S(u,t) P^{\Phi}_{ut} \, du \\ &= \int_0^t P^{\Phi}_{0u} \sum_{n=1}^{\infty} n \sum_{m=n}^{\infty} D_m^S \binom{m}{n} p(u)^n q(u)^{m-n} P^{\Phi}_{ut} \, du \end{aligned}$$

abbreviating $p(u) := G^c((t-u)-)$ and $q(u) := G((t-u)-)$. We further obtain

$$\begin{aligned}
E(Q_t^S) &= \int_0^t P_{0u}^{\Phi} \sum_{n=1}^{\infty} n \sum_{m=n}^{\infty} D_m^S \binom{m}{n} p^n(u) q^{m-n}(u) P_{ut}^{\Phi}\, du \\
&= \int_0^t P_{0u}^{\Phi} \sum_{m=1}^{\infty} \sum_{n=1}^{m} n D_m^S \frac{m!}{n! \cdot (m-n)!} p^n(u) q^{m-n}(u) P_{ut}^{\Phi}\, du \\
&= \int_0^t P_{0u}^{\Phi} \sum_{m=1}^{\infty} m D_m^S p(u) \sum_{n=1}^{m} \frac{(m-1)!}{(n-1)! \cdot (m-n)!} \\
&\quad \times p^{n-1}(u) q^{m-n}(u) P_{ut}^{\Phi}\, du \\
&= \int_0^t P_{0u}^{\Phi} \sum_{m=1}^{\infty} m D_m^S G^c((t-u)-) P_{ut}^{\Phi}\, du
\end{aligned}$$

since

$$\begin{aligned}
&\sum_{n=1}^{m} \frac{(m-1)!}{(n-1)! \cdot (m-n)!} p^{n-1}(u) q^{m-n}(u) \\
&\qquad\qquad = \sum_{n=0}^{m-1} \frac{(m-1)!}{n! \cdot (m-1-n)!} p^n(u) q^{m-1-n}(u)
\end{aligned}$$

sums up to one.
☺

Theorem 9.6 *If condition (9.5) holds and the phase process has stationary distribution π, then the expectation of the additive process of Q^S at any time $t \in I\!R_0^+$ starting in phase equlibrium is*

$$E_\pi(Q_t^S) = \int_{\Phi} \pi(dy) \sum_{n=1}^{\infty} n D_n^S(y, \Phi) \cdot \int_0^t G^c(u-)\, du \tag{9.7}$$

Proof: This follows from the above theorem 9.5. Since $\pi P_{0u}^{\Phi} = \pi$ for all $u \in I\!R_0^+$ and $P_{ut}^{\Phi}(y, \Phi) = 1$ for all $y \in \Phi$ and $u < t \in I\!R_0^+$, we have

$$\begin{aligned}
E_\pi(Q_t^S) &= \int_0^t \int_{\Phi} \pi(dy) \sum_{n=1}^{\infty} n D_n^S(y, \Phi) G^c((t-u)-)\, du \\
&= \int_{\Phi} \pi(dy) \sum_{n=1}^{\infty} n D_n^S(y, \Phi) \cdot \int_0^t G^c(v-)\, dv
\end{aligned}$$

after substituting $v := t - u$.
☺

Now the model shall be extended towards spatially variable service time distributions. To this aim, let $\{S_k : k \in I\!N\}$ with $S_k \in \mathcal{R}$, $S_k \cap S_l = \emptyset$ for $k \neq l$, and $\bigcup_{k=1}^{\infty} S_k = R$ be a partition of the arrival space R. For every $k \in I\!N$, assign a service time distribution function G_k, which is valid on S_k.

Let $S \in \mathcal{R}$ be measurable and $K \in I\!N$ such that $S \subset \bigcup_{k=1}^{K} S_k$. The queue process in S can be described as the convolution of the queue processes in $S \cap S_1, \ldots, S \cap S_K$. Thus, for every $t \in I\!R^+$ the distribution P_t^S of users in subset S at time t can be described in terms of the distributions $P_t^{S \cap S_k}$ of users in the subsets $S \cap S_k$ at time t. This results in

$$P_t^S = *_{k=1}^{K} P_t^{S \cap S_k}$$

Here, the marginal characteristic sequences $\tilde{R}^{S \cap S_k}(t)$ which determine the sequences $P_t^{S \cap S_k}$ for $k \in \{1, \ldots, K\}$ are determined as in equations (9.1) and (9.4). Here, the convolution product $*$ is defined iteratively by $*_{k=1}^{1} A_k := A_1$ and

$$*_{k=1}^{n+1} A_k := (*_{k=1}^{n} A_k) * A_{n+1}$$

for all $n \in I\!N$ and sequences $A_1, \ldots, A_{n+1}$ of kernels on Φ.

1.2 Joint distribution in finitely many subsets

The same method as for spatially variable service time distributions can be applied in order to determine the joint distribution of the queue process in finitely many subsets of the arrival space R at a time $t \in I\!R^+$. Be $n \in I\!N$ and $S_1, \ldots, S_n \in \mathcal{R}$ measurable subsets. The joint distribution $P_t^{S_1,\ldots,S_n}$ of users in service after time t shall be defined as the distribution of the $I\!N_0^n$-valued random variable indicating the number of users in $S_1, \ldots, S_n$ at time t.

Fix a time $t \in I\!R^+$ and subsets $S_1, \ldots, S_n \in \mathcal{R}$ the queue is to be observed at. In any infinitesimal time interval $du = \lim_{h \to 0}]u, u+h]$ with $u \in [0, t[$, batch arrivals occur with rate $\Delta^{S_1,\ldots,S_n} = (D_{i_1,\ldots,i_n}^{S_1,\ldots,S_n} : i \in I\!N_0)$ according to the SMAP arrival process. Given that an arrival of batch size $(m_1, \ldots, m_n) \in I\!N^n$ occured during $]u, u+h]$ in $(S_1, \ldots, S_n)$, the probability of $(i_1, \ldots, i_n) \in \prod_{k=1}^{n} \{0, \ldots, m_k\}$ arrivals still being served in $(S_1, \ldots, S_n)$ at time t is distributed by

$$\prod_{k=1}^{n} \binom{m_k}{i_k} G^c((t-u)-)^{i_k} G((t-u)-)^{m_k - i_k},$$

again denoting $G((t-u)-) := \lim_{h \to 0} G(t-(u+h))$ and $G^c(t) := 1 - G(t)$.

Conditioning upon the batch arrival rates, the total rate of $(i_1, \ldots, i_n)$ arrivals into the family $(S_1, \ldots, S_n)$ of subsets during a time interval du which are still in service at time t is

$$R_{i_1,\dots,i_n}^{S_1,\dots,S_n}(u,t) = \sum_{m_1=i_1}^{\infty} \cdots \sum_{m_n=i_n}^{\infty} D_{m_1,\dots,m_n}^{S_1,\dots,S_n} \times \prod_{k=1}^{n} \binom{m_k}{i_k} G^c((t-u)-)^{i_k} G((t-u)-)^{m_k-i_k}$$

for every $u \in [0,t[$ and $i_1,\dots,i_n \in I\!N_0$. As in the one–dimensional case, the time–dependent (n–dimensional) sequence

$$R^{S_1,\dots,S_n}(u,t) := (R_{i_1,\dots,i_n}^{S_1,\dots,S_n}(u,t) : i_1,\dots,i_n \in I\!N_0)$$

can be interpreted as the marginal characteristic sequence of an inhomogeneous SMAP $H_t^{S_1,\dots,S_n} = (H_{ut}^{S_1,\dots,S_n} : u \le t)$ which at time $u = t$ coincides with the infinite server queue that is to be examined.

With definitions and arguments completely analogous to those in section 1.1, one derives the following result:

Theorem 9.7 *The transient distribution of the marginal queue process of Q in the subsets $S_1,\dots,S_n \in \mathcal{R}$ at time $t \in I\!R_0^+$ is determined by*

$$P_t^{S_1,\dots,S_n} = \sum_{k=0}^{\infty} \int_0^t \int_0^{u_1} \cdots \int_0^{u_{k-1}} \tilde{R}^{S_1,\dots,S_n}(u_1) * \dots * \tilde{R}^{S_1,\dots,S_n}(u_k)\, du_k \dots du_1 \quad (9.8)$$

with $\tilde{R}^{S_1,\dots,S_n}(u) := R^{S_1,\dots,S_n}(t-u,t)$ being independent of $t > u$. The case of zero integrals shall be defined as the n–dimensional sequence $Id = (Id_{i_1,\dots,i_n} : i_1,\dots,i_n \in I\!N_0)$ with entries

$$Id_{i_1,\dots,i_n} := \begin{cases} I, & \sum_{k=1}^{n} i_k = 0 \\ 0, & \text{otherwise} \end{cases}$$

and I (resp. 0) denoting the identity kernel (resp. the zero kernel) on Φ. Furthermore, the convolution sequence $C = A * B$ shall be defined by

$$C_{i_1,\dots,i_n} := \sum_{j_1=0}^{i_1} \cdots \sum_{j_n=0}^{i_n} A_{j_1,\dots,j_n} B_{i_1-j_1,\dots,i_n-j_n}$$

for any two sequences A, B of kernels on Φ.

1.3 Asymptotic Stability

The asymptotic behaviour of the homogeneous $SMAP/G/\infty$ queue already is apparent in formula (9.6) of the expectation kernels. The marginal expectation $E(Q_t^S)$ on any set $S \in \mathcal{R}$ remains finite for $t \to \infty$ if and only if the mean service time $E(G)$ as well as the mean arrival rate are finite. As will be shown in this section, these conditions are, as intuition would suggest, the main ingredients of a stability condition for the homogeneous $SMAP/G/\infty$ queue. After first defining stability for homogeneous queues, the marginal expectations of users in any set $S \in \mathcal{R}$ are given for stable $SMAP/G/\infty$ queues. Finally, a necessary and sufficient stability condition is proven for the case of a finite phase space at the end of this section.

A homogeneous queue $Q = (Q_t : t \in \mathbb{R}^+)$ shall be called **stable** if the asymptotic distributions $p^{S_1,\dots,S_n} := \lim_{t\to\infty} P_t^{S_1,\dots,S_n}$ of the marginal queue processes exist for all $n \in \mathbb{N}$ and $S_1, \dots, S_n \in \mathcal{R}$ and the marginal asymptotic distribution of users in the queue has finite expectation.

Theorem 9.8 *Let Q be stable and denote the asymptotic distribution of the phase process J by π. Then the asymptotic expectation of the marginal process Q^S in a set $S \in \mathcal{R}$ is given by*

$$E(Q^S) := \lim_{t\to\infty} E(Q_t^S) = E(G) \cdot \int_\Phi \pi(dy) \sum_{n=1}^{\infty} n D_n^S(y, \Phi)$$

for every $S \in \mathcal{R}$, independent of the initial distribution.

Proof: This follows immediately from equation (9.6) if one recognizes that with t growing to infinity, the term $G^c((t-u)-)$ tends to zero for the times u during which the phase process is not in equilibrium yet.
☺

In the rest of this section, a stability condition is derived for the case of the phase space Φ being finite, which suffices for most applications. Denote the number of phases by m and the m–dimensional vector with all entries being 1 by 1_m.

Theorem 9.9 *Let Q denote a homogeneous $SMAP/G/\infty$ queue with finite phase space Φ. Assume that the phase process J is irreducible and has stationary distribution π. Then Q is stable if and only if the stability condition*

$$\pi \sum_{n=1}^{\infty} n D_n^R 1_m \cdot E(G) < \infty \tag{9.9}$$

holds.

Proof: First we show necessity. Assume that the queue is stable. Then the asymptotic expectation of the marginal process Q^R is given by

$$E(Q^R) = E(G) \cdot \pi \sum_{n=1}^{\infty} nD_n^R 1_m < \infty$$

according to theorem 9.8. This yields the necessity of condition (9.9).

Sufficiency will be shown by comparison to the $M/G/\infty$ queue. Since Q is stable if and only if Q^R is stable, it suffices to refer to Q^R only. Denote the work load process of Q^R by $W = (W_t : t \in I\!R_0^+)$ and define

$$\gamma_{\max} := -\min_{i \in \Phi} D_0^R(i,i) = \max_{i \in \Phi} |D_0^R(i,i)|$$

as the maximal exit rate over all phases. Define $\tilde{Q}$ as the following $M/G/\infty$ queue. The arrival process of $\tilde{Q}$ shall be a Poisson process with rate $\gamma_{\max}$. Let $i_{\max} \in \Phi$ denote a phase with the highest expectation of the arrival batch size, i.e.

$$\frac{1}{\gamma_{i_{\max}}} \sum_{n=1}^{\infty} nD_n^R(i_{\max}, \Phi) \geq \frac{1}{\gamma_j} \sum_{n=1}^{\infty} nD_n^R(j, \Phi)$$

for all $j \in \Phi$. Since there are no absorbing states, the exit rate $\gamma_{i_{\max}}$ is a positive number. Define the service time distribution of $\tilde{Q}$ by

$$\tilde{G} := \frac{1}{\gamma_{i_{\max}}} \sum_{n=1}^{\infty} D_n^R(i_{\max}, \Phi) G^{*n}$$

denoting the service time distribution of Q by G and the n–fold convolution of G by G^{*n}. The $M/G/\infty$ queue constructed in this way has an arrival rate which is an upper bound of the phase specific arrival rates of Q^R. Furthermore, the batch arrivals of Q^R are interpreted as single arrivals in $\tilde{Q}$ with only one server working off the accrued service requirement of all the users in the batch arrival. Thus the work load $\tilde{W} = (\tilde{W}_t : t \in I\!R_0^+)$ of $\tilde{Q}$ is certainly at least as high as the work load W of Q. Since by Wald's equation

$$E(\tilde{G}) = \frac{1}{\gamma_{i_{\max}}} \sum_{n=1}^{\infty} nD_n^R(i_{\max}, \Phi) \cdot E(G)$$

we have $E(\tilde{G}) < \infty$ by assumption of the stability condition. As $\gamma_{\max} < \infty$ is finite, too, $\tilde{Q}$ is stable. This implies that $\tilde{W}$ is positive recurrent. i.e. the expected duration between the time instants of $\tilde{W}$ reaching the state 0 is finite. Since

$$\tilde{W}_t \geq W_t \geq 0$$

almost certainly for all $t \in \mathbb{R}_0^+$, the work load process W of Q is positive recurrent, too. Hence Q has an asymptotic distribution. The finiteness of the asymptotic expectation of the marginal process Q^R is immediate from theorem 9.8 and the stability condition.
☺

The plausibility of condition (9.9) results from the obvious analogy to the stability condition for the resulting $BMAP/G/\infty$ queue on the set R. Of course, the $SMAP/G/\infty$ queue on $(R, \mathcal{R})$ is stable if and only if the resulting $BMAP/G/\infty$ queue on the set R is stable.

2. General Arrival Rates without User Movements

In mobile communication networks, the typical characteristics of the arrival stream change in time periodically (e.g. over the course of the day). Stochastic models reflect this best by inhomogeneous arrival processes. The concept of SMAPs introduced in chapter 7 does provide for arrival rates varying in time and hence can be used to model this phenomenon. An examination analogous to the homogeneous case yields a solution for the corresponding spatial infinite server queue.

In this section, the same method of analysis as in the preceding section 1 shall be applied to spatial infinite server queues with general arrival rates. User movements will be examined in the next section. After the derivation of the transient distribution, an approximation method for the distribution at arbitrary time instants will be given.

2.1 Transient Distribution

The infinite server queue fed by an inhomogeneous SMAP can be analyzed analogously to the homogeneous case in section 1. Let

$$Q = \left(Q_t^{S_1;\dots,S_n} : t \in \mathbb{R}_0^+, n \in \mathbb{N}, S_1;\dots,S_n \in \mathcal{R}\right)$$

denote a spatial queue with general SMAP input (N, J), general service time distribution G and infinitely many servers. The arrival process (N, J) shall be defined by its time–dependent marginal characteristic sequences

$$\left(\Delta^{S_1;\dots,S_n}(t) : t \in \mathbb{R}_0^+, n \in \mathbb{N}, S_1;\dots,S_n \in \mathcal{R}\right)$$

Again, the reason for restriction to a single subset and spatially constant service time distributions is ease of notation. The generalizations towards spatially variable service times and joint distributions in finitely many subsets are straightforward in the same way as in section 1. Be $t \in \mathbb{R}^+$ the time instant and $S \in \mathcal{R}$ the subset the queue is observed at. Now the arrival rates at times $s \in [0, t]$ are not constant anymore, but depend on the time instant s.

The total rate of i arrivals during a time interval $du = \lim_{h\to 0}]u, u+h]$ is given by

$$R_i^S(u,t) = \sum_{m=i}^{\infty} D_m^S(u) \binom{m}{i} G^c((t-u)-)^i G((t-u)-)^{m-i}$$

for every $u \in [0,t[$ and $i \in I\!N_0$. As in the case of homogeneous arrival rates, the sequence $R^S(u,t) = (R_i^S(u,t) : i \in I\!N_0)$ can be interpreted as the time–dependent sequence of transition rates of an inhomogeneous SMAP $H^{(t;S)} = (H^{(t;S)} : u \le t)$ which at time $u = t$ coincides with the infinite server queue that is to be examined. The same arguments as in section 1.1 lead to

Theorem 9.10 *The transient distribution of the marginal queue process Q^S in the subset $S \in \mathcal{R}$ at time $t \in I\!R_0^+$ is determined by*

$$P_t^S = \sum_{k=0}^{\infty} \underbrace{\int_0^t \int_0^{u_k} \dots \int_0^{u_2}}_{k \text{ integrals}} R^S(u_1,t) * \dots * R^S(u_k,t)\, du_1 \dots du_k \qquad (9.10)$$

defining the case of zero integrals as the sequence $Id = (I, 0, 0, \dots)$ with I denoting the identity kernel and 0 the zero kernel on Φ, and further defining the convolution sequence $C = A * B$ by

$$C_n := \sum_{i=0}^{n} A_i B_{n-i}$$

for any two sequences A, B of kernels on Φ.

Unfortunately, a unification of the above formula analogous to formula (9.3) cannot be achieved in general. This is due to the fact that

$$R^S(t-u,t) = \sum_{m=i}^{\infty} D_m^S(t-u) \binom{m}{i} G^c(u-)^i G(u-)^{m-i}$$

still depends on t if the arrival rates are inhomogeneous.

On the other hand, it is possible to extend the theory towards time–dependent service time distribution functions. This would lead to rates of the form

$$R_i^S(u,t) = \sum_{m=i}^{\infty} D_m^S(u) \binom{m}{i} G_u^c((t-u)-)^i G_u((t-u)-)^{m-i}$$

for every $u \in [0,t[$ and $i \in I\!N_0$, with G_u denoting the service time distribution function which holds for users arriving during the infinitesimal time interval du.

2.2 An Approximation

This section shows an approximation method which can be used in order to determine the distribution of the queue process at an arbitrary time $t \in I\!R_0^+$ without needing to integrate over the whole range $[0, t[$. The approximation regards only arrivals up to a certain time distance into the past. All arrivals which have occured before are neglected. Thus, this method approximates the service time distribution by a distribution with cut tail.

Conditions for this approximation are the standard assumptions of finite mean service time and finite arrival rates. First, it is shown that the kernels of the marginal characteristic sequence of $H^{(t;S)}$ are uniformly bounded. Then, this result is used to estimate the approximation error. Assume in this chapter that the mean service time

$$E(G) = \int_0^\infty G^c(u)\, du < \infty \tag{9.11}$$

is finite. Further assume that the arrival rate in the whole arrival space R

$$\sum_{m=1}^{\infty} m D_m^R(t)(y, \Phi) < M < \infty$$

is finite for all $t \in I\!R^+$ and $y \in \Phi$. The next result gives the decisive bound for the approximation following.

Lemma 9.11 *For every $\varepsilon > 0$, there is a time distance $T(\varepsilon) \in I\!R^+$ such that the inequality*

$$\sup_{y \in \Phi} \left| \int_0^s R_n^S(u, t)(y, \Phi)\, du \right| < M \cdot \varepsilon$$

holds for all $n \in I\!N_0$ and $s \leq t - T(\varepsilon)$.

Proof: Choose any $\varepsilon > 0$ and abbreviate $\|L\| := \sup_{y \in \Phi} |L(y, \Phi)|$ for any kernel L on Φ. By assumption (9.11), there is a $T(\varepsilon) \in I\!R^+$ such that

$$\int_{T(\varepsilon)}^\infty G^c(u-)\, du < \varepsilon$$

Fix any $y \in \Phi$ and $s \leq t - T(\varepsilon)$. Since all elements of $R_n^S(u, t)$ are pesitive for all $u \in [0, t]$ and $n \geq 1$, we have

$$\left\| \int_0^s \sum_{n=1}^\infty R_n^S(u, t)\, du \right\| \leq \left\| \int_0^s \sum_{n=1}^\infty n R_n^S(u, t)\, du \right\|$$

Analogously to the proof of theorem 9.5, one can show that

$$\int_0^s \sum_{n=1}^{\infty} nR_n^S(u,t)\, du = \int_0^s \sum_{n=1}^{\infty} nD_n^S(u)\, G^c((t-u)-)\, du$$

Because $\sum_{n=1}^{\infty} nD_n^S(u) \leq \sum_{n=1}^{\infty} nD_n^R(u)$ for all $S \in \mathcal{R}$ and $u \in I\!R_0^+$, the estimation

$$\begin{aligned}
\left\| \int_0^s R_k^S(u,t)\, du \right\| &\leq \left\| \int_0^s \sum_{n=1}^{\infty} R_n^S(u,t)\, du \right\| \\
&\leq \left\| \int_0^s \sum_{n=1}^{\infty} nD_n^R(u)\, G^c((t-u)-)\, du \right\| \\
&\leq \int_0^s \left\| \sum_{n=1}^{\infty} nD_n^R(u) \right\| G^c((t-u)-)\, du \\
&< M \cdot \int_0^s G^c((t-u)-)\, du
\end{aligned}$$

holds for all $k \in I\!N$. Substituting $v := t-u$ leads to

$$\begin{aligned}
\left\| \int_0^s R_k^S(u,t)\, du \right\| &< M \cdot \int_{t-s}^{t} G^c(v-)\, dv \leq M \cdot \int_{T(\varepsilon)}^{\infty} G^c(v-)\, dv \\
&< M \cdot \varepsilon
\end{aligned}$$

Since

$$\int_0^s \sum_{n=1}^{\infty} R_n^S(u,t)(y,\Phi) du = -\int_0^s R_0^S(u,t)(y,\Phi) du$$

for all $s \in I\!R_0^+$, the proof is complete.
☺

Using this bound, we can prove the following approximation:

Theorem 9.12 *At any time $t \in I\!R^+$ and for every $y \in \Phi$, the distribution of the queue process can be approximated by*

$$P_t^S(n)(y,\Phi) = P_{0t}^S(n)(y,\Phi) \approx P_{t-T(\varepsilon),t}^S(n)(y,\Phi)$$

for all $n \in I\!N_0$ and $y \in \Phi$. The approximation error is at most

$$\left| P_t^S(n)(y,\Phi) - P_{t-T(\varepsilon),t}^S(n)(y,\Phi) \right| < M \cdot \varepsilon$$

independently of $n \in I\!N_0$ and $y \in \Phi$.

Proof: For every $k \in I\!N$, the difference between the respective k-fold integrals appearing in formula (9.10) is

$$\int_0^t \cdots \int_0^{u_{k-1}} R^S(u_k,t) * \ldots * R^S(u_1,t)\, du_k \ldots du_1$$

$$- \int_{t-T(\varepsilon)}^t \cdots \int_{t-T(\varepsilon)}^{u_{k-1}} R^S(u_k,t) * \ldots * R^S(u_1,t)\, du_k \ldots du_1$$

$$= \int_0^t \cdots \int_0^{u_{k-2}} \int_0^{t-T(\varepsilon)} R^S(u_k,t) * \ldots * R^S(u_1,t)\, du_k \ldots du_1$$

$$= \int_0^{t-T(\varepsilon)} R^S(u_k,t)\, du_k$$

$$* \int_0^t \cdots \int_0^{u_{k-2}} R^S(u_{k-1},t) * \ldots * R^S(u_1,t)\, du_{k-1} \ldots du_1$$

Summing up over all $k \in I\!N_0$ yields

$$P_t^S(\cdot) - P_{t-T(\varepsilon),t}^S(\cdot)$$

$$= \int_0^{t-T(\varepsilon)} R^S(u,t)\, du$$

$$* \sum_{k=0}^{\infty} \int_0^t \cdots \int_0^{u_{k-1}} R^S(u_k,t) * \ldots * R^S(u_1,t)\, du_k \ldots du_1$$

$$= \int_0^{t-T(\varepsilon)} R^S(u,t)\, du * P_t^S$$

Hence, for any $y \in \Phi$ and $n \in I\!N_0$ we have

$$|P_t^S(n)(y,\Phi) - P_{t-T(\varepsilon),t}^S(n)(y,\Phi)|$$

$$= \left| \sum_{j=0}^{n} \int_{u=0}^{t-T(\varepsilon)} \int_{\Phi} R_j^S(u)(y,dz)du\; P_t^S(n-j)(z,\Phi) \right|$$

$$\le \max_{j=0,\ldots,n} \left| \int_0^{t-T(\varepsilon)} R_j^S(u)(y,\Phi)\, du \right|$$

$$< M \cdot \varepsilon$$

according to lemma 9.11, since the probability $\sum_{j=0}^{n} \left| P_t^S(n-j)(z,\Phi) \right|$ is bounded by one for every $z \in \Phi$.
☺

Concluding this section, we give two immediate applications of the above approximation for a computation of the asymptotic distribution of the homogeneous spatial infinite server queue.

Corollary 9.13 *Let Q denote a stable homogeneous $SMAP/G/\infty$ queue with service time distribution G. If the service time is bounded, i.e. if there is a time T such that $G(T) = 1$, then the asymptotic distribution of Q is given by*

$$p^{S_1,\ldots,S_n} = \sum_{k=0}^{\infty} \int_0^T \int_0^{u_1} \cdots \int_0^{u_{k-1}} \int_\Phi d\pi(y) \times \left(\tilde{R}^{S_1,\ldots,S_n}(u_1) * \ldots * \tilde{R}^{S_1,\ldots,S_n}(u_k)\right)(y, \Phi)\, du_k \ldots du_1$$

for all $n \in I\!N$ and $S_1, \ldots, S_n \in \mathcal{R}$, with rate kernels $\tilde{R}^{S_1,\ldots,S_n}(u)$ as defined in section 1.2.

Corollary 9.14 *Let Q denote a stable homogeneous $SMAP/G/\infty$ queue with service time distribution G. Choose any $\varepsilon > 0$ and determine $T(\varepsilon)$ as in lemma 9.11. Then an approximation of the asymptotic distribution of Q^S is given by*

$$p^S \approx \sum_{k=0}^{\infty} \int_0^{T(\varepsilon)} \int_0^{u_1} \cdots \int_0^{u_{k-1}} \int_\Phi d\pi(y) \times \left(\tilde{R}^S(u_1) * \ldots * \tilde{R}^S(u_k)\right)(y, \Phi)\, du_k \ldots du_1$$

for any set $S \in \mathcal{R}$, again with rate kernels $\tilde{R}^S(u)$ as defined in section 1.2. The approximation error is uniformly bounded by $M \cdot \varepsilon$, with M denoting a bound for the arrival rate in R.

Example 9.15 As in example 9.4, assume that the service times are deterministically s. Further assume that the phase process is asymptotic with asymptotic distribution π. Then the asymptotic distribution of Q^S under any initial distribution ν is given by

$$p^S = \lim_{t\to\infty} \int d\nu(y) P_t^S(y, \Phi) = \left(\lim_{t\to\infty} \int d\nu(y) P_{0,t}^\Phi\right) e^{\Delta^S \cdot s}(y, \Phi) = \int d\pi(y) e^{\Delta^S \cdot s}(y, \Phi)$$

for all $S \in \mathcal{R}$.

3. General Arrival Rates with User Movements

User movements shall be modeled following the idea presented in Çinlar [44], pp.112f. Divide the space R into a finite partition $(S_1, \ldots, S_K) \in (\mathcal{R})^K$, such that $K \in \mathbb{N}$, $\bigcup_{i=1}^{K} S_i = R$ and $S_i \cap S_j = \emptyset$ for $i \neq j$. Each set of this partition shall represent a spatial unit which is homogeneous in motion. Thus for times $s < t \in \mathbb{R}^+$, subsets $S_i, S \in \mathcal{R}$ and $n, k \in \mathbb{N}_0$, one can define a probability $P_{nk}^{S_i,S}(s,t)$ that under the condition that n users arrive in S_i at time s, k of them will be located in S at time t. At the time a user has finished using the network (i.e. its service time is over), it shall disappear and take on the position $v \notin R$. This implies $P_{nn}^{S_i,R\cup\{v\}}(s,t) = 1$ for all $S_i \in \mathcal{R}$, $s < t \in \mathbb{R}^+$ and $n \in \mathbb{N}_0$. Further, spatially variable service times can be subsumed by $G_i(t) = P_{1,0}^{S_i,R}(0,t)$ or $\binom{m}{k} G_i^c(t)^k G_i(t)^{m-k} = P_{mk}^{S_i,R}(0,t)$ for batch arrivals.

The transient distribution of the queue process can be calculated as a convolution over all $i \in \{1, \ldots, K\}$ of the distributions of users which arrived in the set S_i until time t and assume a location in S at time t. As in equation (9.1), the total rate of k users arriving in S_i during the infinitesimal time interval $du = \lim_{h\to 0}]u, u+h]$ and being located in S at time t is determined by

$$R_k^{S_i,S}(u,t) = \sum_{m=k}^{\infty} D_m^{S_i}(u) P_{mk}^{S_i,S}(u,t)$$

for every $u \in [0,t[$, $k \in \mathbb{N}_0$ and $S_i \in \mathcal{R}$. Defining the sequence of matrices $R^{S_i,S}(u,t) := (R_k^{S_i,S}(u,t) : k \in \mathbb{N}_0)$ for every $u \in [0,t[$, the above rates determine the distributions $P_t^{S_i,S}$ of users arriving in S_i and being located in S at time t. The same arguments leading to formula (9.2) yield the expression

$$P_t^{S_i,S} = \sum_{k=0}^{\infty} \underbrace{\int_0^t \int_0^{u_k} \cdots \int_0^{u_2}}_{\text{k integrals}} R^{S_i,S}(u_1,t) * \ldots * R^{S_i,S}(u_k,t)\, du_1 \ldots du_k$$

The distribution of the queue as a convolution of these single distributions is given as

$$P_t^S = *_{i=1}^{K}\, P_t^{S_i,S}$$

for all $t \in \mathbb{R}^+$ and $S \in \mathcal{R}$. An approximation of this expression is possible by the same idea as in section 2.2. First let $P_{st}^{S_i,S}(n)$ denote the (sub–stochastic) kernel of n users arriving in S_i during time $]s,t]$ and being located in S at time t. Thus we can write $P_t^{S_i,S} = P_{0t}^{S_i,S}$. Further, let $P_{st}^S(n)$ the (sub–stochastic) kernel of n users arriving anywhere during time $]s,t]$ and being located in S at time t. Likewise, we write $P_t^S = P_{0t}^S$. In order to obtain a bound for the approximation error, we need the following

Lemma 9.16 *Assume that the mean service time is finite, i.e. that equation (9.11) holds. Be $\varepsilon > 0$ and determine $T(\varepsilon)$ as in theorem 9.12. Then the*

inequalitiy

$$\sup_{y\in\Phi}\left|\prod_{i=1}^{K} P_{0t}^{S_i,S}(n_i)(y,\Phi) - \prod_{i=1}^{K} P_{t-T(\varepsilon),t}^{S_i,S}(n_i)(y,\Phi)\right| < K\cdot M\varepsilon$$

holds for all $n_1,\ldots,n_K \in I\!N_0$.

Proof: This is proven by induction. For $K = 1$ the statement is proven in the same way as in theorem 9.12. Assume the statement is true for any $K \geq 1$. Choose any $y \in \Phi$. To shorten the notation, define $p_i := P_{0t}^{S_i,S}(n_i)(y,\Phi)$ and $\tilde{p}_i := P_{t-T(\varepsilon),t}^{S_i,S}(n_i)(y,\Phi)$. As probabilities, all p_i and $\tilde{p}_i$ have values in $[0,1]$. Then

$$\begin{aligned}\left|\prod_{i=1}^{K} p_i - \prod_{i=1}^{K}\tilde{p}_i\right| &= \left|\prod_{i=1}^{K-1} p_i(p_K - \tilde{p}_K) + \left(\prod_{i=1}^{K-1} p_i - \prod_{i=1}^{K-1}\tilde{p}_i\right)\tilde{p}_K\right| \\ &< \prod_{i=1}^{K-1} p_i\cdot M\varepsilon + (K-1)\cdot M\varepsilon\cdot\tilde{p}_K\end{aligned}$$

by induction hypothesis. Since $\prod_{i=1}^{K-1} p_i \leq 1$ and $\tilde{p}_K \leq 1$, the induction step is proven.
☺

Since the combinatorial possibilities of creating $n \in I\!N_0$ users out of K different sources grows with n, an error for the approximation of $P_t^S(n)$ can only be given dependent on the number n of users. This is the content of

Theorem 9.17 *The distribution of the queue process can be approximated by*

$$P_{0t}^{S}(n)(y,\Phi) \approx P_{t-T(\varepsilon),t}^{S}(n)(y,\Phi)$$

for every $n \in I\!N_0$ *and* $y \in \Phi$. *The approximation error is at most*

$$\left|P_{0t}^{S}(n)(y,\Phi) - P_{t-T(\varepsilon),t}^{S}(n)(y,\Phi)\right| < \binom{K+n-1}{K-1}\cdot K\cdot M\varepsilon$$

inependently of $y \in \Phi$.

Proof: Using the abbreviations $p_i(n_i) := P_{0t}^{S_i,S}(n_i)(y,\Phi)$ and $\tilde{p}_i(n_i) := P_{t-T(\varepsilon),t}^{S_i,S}(n_i)(y,\Phi)$ as in the previous proof, the independence of the user movements leads to the expressions

$$P_{0t}^{S}(n)(y,\Phi) = \sum_{n_1+\ldots+n_K=n}\prod_{i=1}^{K} p_i(n_i)$$

and

$$P^S_{t-T(\varepsilon),t}(n)(y,\Phi) = \sum_{n_1+\ldots+n_K=n} \prod_{i=1}^{K} \tilde{p}_i(n_i)$$

Now the approximation error is bounded by

$$\left| \sum_{n_1+\ldots+n_K=n} \prod_{i=1}^{K} p_i(n_i) - \sum_{n_1+\ldots+n_K=n} \prod_{i=1}^{K} \tilde{p}_i(n_i) \right|$$
$$= \left| \sum_{n_1+\ldots+n_K=n} \left(\prod_{k=1}^{K} p_i(n_i) - \prod_{k=1}^{K} \tilde{p}_i(n_i) \right) \right| < \sum_{n_1+\ldots+n_K=n} K \cdot M\varepsilon$$

according to lemma 9.16. As can be proven by induction on K, the number of combinations of summing up $n_1 + \ldots + n_K = n$ is $\binom{K+n-1}{K-1}$.
☺

4. Application: Planning a Mobile Communication Network

The results which have been obtained in this chapter can be applied to support the planning procedure of mobile communication networks. Assume that an area R is to be covered with base stations such as to provide network service for all users in it. This shall be achieved with a maximum outage probability of ε, which means that a user at any location in R, who wants to use the network, will find it busy with a probability of at most ε.

Then the analysis of spatial infinite server queues can help to derive a capacity function $C_\varepsilon : \mathcal{R} \times I\!R^+ \to I\!N_0$ which yields for every subset $S \in \mathcal{R}$ and every time $t \in I\!R^+$ a number $C_\varepsilon(S,t)$ such that if the network has the capacity of serving $C_\varepsilon(S,t)$ users in S at time t, then the quality of service with respect to the given maximum outage probability ε is guaranteed. Hence, in order to build a network that can guarantee the maximum outage probability ε, it suffices to provide service capacity according to the function C_ε. In practical applications, a finite partition $(S_1, \ldots, S_K)$ of R will be fine enough to answer all relevant questions concerning network planning. This implies that $\mathcal{R}$ may be chosen as finite, too, which greatly simplifies the computation of the function C_ε.

In order to determine C_ε, one can proceed as follows. Define a spatial $SMAP/G/\infty$ queue with arrival space $(R, \mathcal{R})$ on an appropriate level regarding user movements and time–dependence of arrival rates. Then either the stationary distribution (in the homogeneous case) or the distributions at a given time of day (in the case of periodic arrival rates with period length being a day) can be represented by a set of functions $P_t : \mathcal{R} \times I\!N_0 \to [0,1]$ which yields for a given set $S \in \mathcal{R}$ the probability $P_t(S,n)$ of n users being served in S at day time t. For the homogeneous case, we have a function P instead of a set $(P_t : t \in [0, T])$ of functions which depend on the day time.

These are the distributions for an infinite server queue, which means that the probabilities were derived under the assumption that every user can be served. Hence the load for any real network with some positive outage probability would be lower. It thus suffices to determine the capacity function C_ε for the infinite server queue in order to find a sufficient capacity function for any real network. Since the outage probability ε usually will be chosen very small, the approximation by an infinite server queue seems reasonably sharp.

Now the capacity function for the infinite server queue can easily be determined as

$$C_\varepsilon(S,t) := \min\left\{ n \in I\!N_0 : \sum_{k=n+1}^{\infty} P_t(S,k) < \varepsilon \right\}$$

for all $t \in I\!R^+$ and $S \in \mathcal{R}$. This means that the value $C_\varepsilon(S,t)$ of the capacity function at a subset $S \in \mathcal{R}$ at a time t is given by the lowest number n of users such that the probability of more than n users being served in S would be smaller than ε for the respective infinite server queue. Since the load of any real network is lower than the load for the infinite server queue, the outage probability of a network satisfying this capacity function can be guaranteed to stay smaller than the given threshold ε.

Chapter 10

MODEL FITTING FOR A CLASS OF SMAPS

In order to apply the concept of SMAPs and their respective queues, it is necessary to estimate the arrival parameters from real data streams. In general, this problem has not been solved yet. The problem of estimating parameters is more complex for SMAPs than for Batch Markovian Arrival Processes (BMAPs, see chapters 3 and 6), since homogeneous SMAPs generalize BMAPs towards a larger phase space and - most importantly - spatial batch arrivals. However, for certain subclasses of SMAPs estimation procedures can be derived.

In this chapter, an estimation procedure is introduced for the subclass of homogeneous SMAPs which have only finitely many phases and a finite σ–algebra $\mathcal{R}$ on the arrival space R. For this class of SMAPs, the counting function of an arrival is reduced to a vector $C \in \mathbb{N}_0^K$ with $K = |\mathcal{R}|$ denoting the number of elements in $\mathcal{R}$. Since BMAPs belong to this subclass of SMAPs as a special case, the proposed estimation procedure yields a new result for BMAPs, too. This is explicated in chapter 6.

The estimation procedure introduced in this chapter requires the measurement of the arrival time instants and the counting functions of the arrivals. Let $(T_n : n \in \{0, \ldots, N\})$ denote the empirical arrival instants and $(C_n : n \in \{0, \ldots, N\})$ the respective empirical counting functions of the spatial arrivals at times $(T_n : n \in \{0, \ldots, N\})$. Define as usual $T_0 := 0$.

The estimation works in three steps, each of which uses classical statistical methods. First, the empirical interarrival times are used to estimate the number m of phases and the matrix D_0^R. If we first neglect the correlations between the inter–arrival times, these can be interpreted as a sample of a phase–type distribution and hence D_0^R can be estimated by an EM-algorithm (see Dempster et al. [50] or for this special case Asmussen et al. [7]). If the number of phases is unknown, then the above maximum likelihood estimation can be repeated with increasing phase spaces until the likelihood gain does not exceed a certain threshhold (cf. Jewell [72]).

In a second step, we want to re–introduce the correlations between the inter–arrival times. To this aim, for every empirical arrival instant T_n the probability distribution of being in a certain phase immediately before and after this instant is estimated using discriminant analysis (see Titterington et al. [122]). Thus the correlation structure is estimated from the estimate $\hat{D}_0$ derived in the first step and the empirical inter–arrival times.

In the last step, the derived estimators of the first two steps are used in order to calculate the empirical estimator for the generator matrix. This is done according to standard estimators for Markov chains (see Anderson and Goodman [2]).

1. Arrival Rates

Let $(z_n := T_n - T_{n-1} : n \in \{1, \ldots, N\})$ denote the empirical interarrival times. Assume first that the number of phases is known and denoted by m. According to theorem 7.15, the interarrival times of an SMAP are distributed phase–type with generator D_0^R. Hence if we assume the phase distribution after the arrival instants T_n to be constant, say $\pi = (\pi_1, \ldots, \pi_m)$, the $(z_n : n \in \{1, \ldots, N\})$ can be interpreted as a sample of a phase–type distribution with density

$$z(t) = \pi e^{D_0^R t} \eta^R$$

for $t \in \mathbb{R}_+$. Here, $\eta^R := -D_0^R 1_m$ is the so–called exit vector of the phase–type distribution with representation (π, D_0^R). For an introduction to phase–type distributions see Neuts [96], chapter 2.

For ease of notation, we write in this and the following section D_0 instead of D_0^R and η instead of η^R. Denote the ith unit column vector by e_i and its transposition (i.e. the respective row vector) by e_i^T.

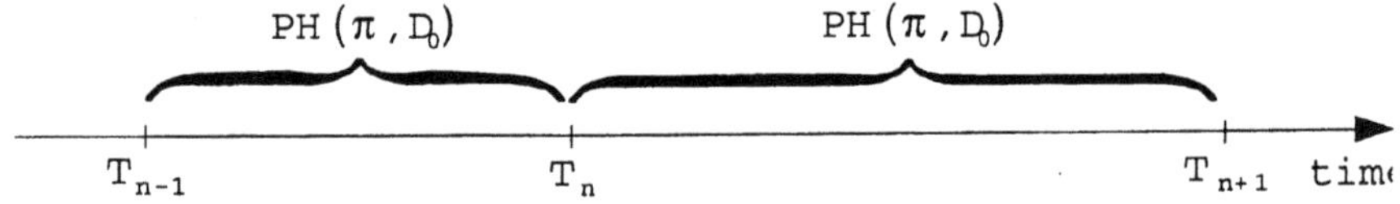

Neglecting the correlations between inter–arrival times

If m is known, there is a maximum likelihood estimator for π and D_0^R. The solution of the estimating equations can be approximated iteratively by an EM algorithm (cf. Dempster et al. [50] or McLachlan and Krishnan [85]), which was derived for this special case by Asmussen et al. [7] and proceeds as follows.

Starting from an intuitive first estimate $(\pi^{(1)}, D_0^{(1)})$ of the representation of the phase–type distribution, the recursions

$$\pi_i^{(k+1)} = \frac{1}{N}\sum_{n=1}^{N} \frac{\pi_i^{(k)} b_i^{(k)}(z_n)}{\pi^{(k)} b^{(k)}(z_n)}$$

$$D_{0;ij}^{(k+1)} = \sum_{n=1}^{N} \frac{D_{0;ij}^{(k)} c_{ij}^{(k)}(z_n)}{\pi^{(k)} b^{(k)}(z_n)} \Big/ \sum_{i=1}^{N} \frac{c_{ii}^{(k)}(z_n)}{\pi^{(k)} b^{(k)}(z_n)}$$

for $i \neq j$ and

$$\eta_i^{(k+1)} = \sum_{n=1}^{N} \frac{\eta_i^{(k)} a_i^{(k)}(z_n)}{\pi^{(k)} b^{(k)}(z_n)} \Big/ \sum_{i=1}^{N} \frac{c_{ii}^{(k)}(z_n)}{\pi^{(k)} b^{(k)}(z_n)}$$

along with the relation

$$D_{0;ii}^{(k+1)} = -\eta_i^{(k+1)} - \sum_{j=1, j\neq i}^{m} D_{0;ij}^{(k+1)}$$

and the definitions

$$a^{(k)}(z_n) := \pi^{(k)} e^{D_0^{(k)} z_n}$$

$$b^{(k)}(z_n) := e^{D_0^{(k)} z_n} \eta^{(k)}$$

$$c_{ij}^{(k)}(z_n) := \int_0^{z_n} \pi^{(k)} e^{D_0^{(k)} u} e_i e_j^T e^{D_0^{(k)}(z_n - u)} \eta \, du$$

for $i, j \in \{1, \ldots, m\}$ und $k \in I\!N$ lead to monotonically increasing likelihoods.

In Asmussen et al. [7], it is proposed to compute the values of $a^{(k)}(z_n)$, $b^{(k)}(z_n)$ and $c_{ij}^{(k)}(z_n)$ numerically as the solution to a linear system of homogeneous differential equations. According to their paper, this is possible achieving high precision by standard methods. The implementation by Gilbert [63] shows that, too.

The convergence of the EM algorithm is examined in Dempster et al. [50], chapter 3, as well as in McLachlan and Krishnan [85]. Improvements of the convergence rates are given in Meng and Dyk [88] and Jamshidian, Jennrich [71]. Titterington [121] gives an alternative recursion which is faster but not as stable.

A satisfying standard procedure for estimating the number m of phases has not been found yet. A feasible method without a prior estimation of m is proposed in Jewell [72]. Denote the maximum likelihood estimators of the representation of the phase–type distribution (as approximated by the EM algorithm described above) for an assumed number m_k of phases by $(\hat{\pi}(k), \hat{D}_0(k))$. Further denote the resulting estimate of the exit vector by $\hat{\eta}(k) := -\hat{D}_0(k)1_{m_k}$.

Estimating the parameters by the above method for increasing m_k and stopping as soon as the likelihood gain

$$\prod_{n=1}^{N} \hat{\pi}(k+1) e^{\hat{D}_0(k+1) z_n} \hat{\eta}(k+1) \Big/ \prod_{n=1}^{N} \hat{\pi}(k) e^{\hat{D}_0(k) z_n} \hat{\eta}(k)$$

is smaller than a threshold value leads to a reasonable model fitting. Since the adaptation of the model increases with the assumed number of phases, the likelihood gain is always positive. The threshold value reflects the limit of accuracy beyond which the gain in model adaptation is not worth the additional computation time.

2. Phases at Arrival Instants

Using the estimates $(\hat{\pi}, \hat{D}_0)$ from the last section, the distribution of the non–observable phases at times $(T_n : n \in \mathbb{N})$ can be estimated using discriminant analysis in a standard way (cf. Titterington et al. [122], pp.168f).

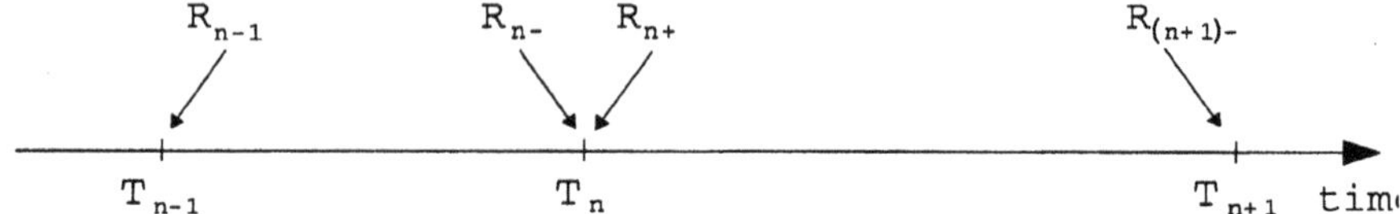

Re–introducing correlations via estimation of phases

For a given empirical arrival instant T_n, let $Z_n := T_n - T_{n-1}$ and R_{n-1} denote the random variables of the last interarrival time and of the phase immediately after the last arrival instant T_{n-1}, respectively. Let $P(Z_n, R_{n-1})$ denote their joint distribution and $P(Z_n|R_{n-1})$, $P(R_{n-1}|Z_n)$ the conditional distributions.

Since $(Z_n : n \in \{1, \ldots, N\})$ is given empirically by the time series $(T_n : n \in \{0, \ldots, N\})$, it suffices to estimate the distribution $P(R_{n-1}|Z_n = z_n)$ for every $n \in \{1, \ldots, N-1\}$. This distribution is discrete, because by assumption there are only finitely many phases.

Bayes' formula yields for $j \in \{1, \ldots, m\}$ and $n \in \{1, \ldots, N\}$

$$P(R_{n-1} = j|Z_n = z_n) = \frac{P(Z_n = z_n|R_{n-1} = j) \cdot P(R_{n-1} = j)}{\sum_{i=1}^{m} P(Z_n = z_n|R_{n-1} = i) \cdot P(R_{n-1} = i)}$$

Since the interarrival times Z_n are distributed phase–type, the expressions $P(Z_n|R_{n-1})$ exist as conditional densities with respect to the Lebesgue measure on $\mathbb{R}$.

The expressions on the right hand can be estimated via the estimated parameters $(\hat{\pi}, \hat{D}_0)$ and the resulting vector $\hat{\eta}$. Hence, for every arrival instant

T_{n-1} the estimator for the conditional distribution of the phase at T_{n-1} given the interarrival time $Z_n = T_n - T_{n-1}$ is given by

$$\hat{P}(R_{n-1} = i | Z_n = z_n) = \frac{e_i^T e^{\hat{D}_0 z_n} \hat{\eta} \cdot \hat{\pi}_i}{\sum_{j=1}^m e_j^T e^{\hat{D}_0 z_n} \hat{\eta} \cdot \hat{\pi}_j} = \frac{\hat{\pi}_i \cdot e_i^T e^{\hat{D}_0 z_n} \hat{\eta}}{\hat{\pi} e^{\hat{D}_0 z_n} \hat{\eta}}$$

for every $i \in \{1, \ldots, m\}$ and $n \in \{1, \ldots, N\}$.

Furthermore, it will be necessary to estimate the phase immediately before an arrival instant. This can be done by the same method. Denote the respective conditional distribution by $P(R_{n-}|Z_n)$. Then we get the estimation

$$\hat{P}(R_{n-} = i | Z_n = z_n) = \frac{\hat{\pi} e^{\hat{D}_0 z_n} e_i \cdot \hat{\eta}_i}{\sum_{j=1}^m \hat{\pi} e^{\hat{D}_0 z_n} e_j \cdot \hat{\eta}_j} = \frac{\hat{\pi} e^{\hat{D}_0 z_n} e_i \cdot \hat{\eta}_i}{\hat{\pi} e^{\hat{D}_0 z_n} \hat{\eta}}$$

for every $i \in \{1, \ldots, m\}$ and $n \in \{1, \ldots, N\}$.

3. Generator Matrix

Since the phase space $\Phi = \{1, \ldots, m\}$ as well as the σ–algebra $\mathcal{R}$ is finite, the infinitesimal transition rates are determined by the values for

$$D(i, j, C) = \hat{\eta}_i \cdot p_i(j, C)$$

for every $i, j \in \{1, \ldots, m\}$ and $C \in \mathbb{N}_0^K$, $C \neq \mathbf{0}$.

Denote the zero vector in $\mathbb{N}_0^K$ by $\mathbf{0}$. In order to complete the estimation of the generator matrix, it suffices to estimate the parameters $p_i(j, C)$ for $C \neq \mathbf{0}$. Without any further assumptions regarding these, use of the empirical estimator is standard. This is given in Anderson and Goodman [2]. Since in the present statistical model the phase process is hidden, the phase at each arrival instant cannot be observed but must be estimated. For this, the results of the last section are used.

For every $k \in \{1, \ldots, N-1\}$, let R_k and R_{k-} denote the random variable of the (non-observable) phase at the empirical arrival instant T_k and immediately before it, respectively. In the last section, estimators for the conditional distributions of the R_k and R_{k-} were given. Let δ denote the Kronecker function for testing equality of counting functions and remember that C_k denotes the empirical counting function of the kth spatial arrival. Define

$$N_i(j, C) := \sum_{k=1}^{N-1} \hat{P}(R_{k-} = i | Z_k = z_k) \cdot \delta_{C_k, C} \cdot \hat{P}(R_k = j | Z_{k+1} = z_{k+1})$$

and

$$N_i := \sum_{k=1}^{N-1} \hat{P}(R_{k-} = i | Z_k = z_k) = \sum_{C \in \mathbb{N}_0^K} \sum_{j-1}^{m} N_i(j, C)$$

for $C \in \mathbb{N}_0^K$ and $i, j \in \{1, \dots, m\}$. In an SMAP, the random variables R_{k-} and R_k are dependent in any non–trivial case. Since in the present model the phases are non–observable, this dependence can only be reflected by conditioning on the consecutive empirical interarrival times z_k and z_{k+1}.

The empirical estimator for $p_i(j, C)$ can now be computed as

$$\hat{p}_i(j, C) := \frac{N_i(j, C)}{N_i}$$

for every $C \in \mathbb{N}_0^K \setminus \{\mathbf{0}\}$ and $i, j \in \{1, \dots, m\}$. This completes the last step of the estimation procedure for SMAPs.

Remark 10.1 For applications in the modelling of mobile communication networks, it usually suffices to set $\mathcal{R} := \sigma(\{S_1, \dots, S_k\})$ with $S_1, \dots, S_k$ being a finite partition of R, i.e. $R = \sum_{i=1}^k S_i$ and $S_i \cap S_j = \emptyset$ for $i \neq j$. As the discrete σ-algebra over a finite set of subsets, $\mathcal{R}$ will then be finite, too. In such an application field, the subsets S_i describe entities with roughly homogeneous call behaviour like big buildings, housing blocks, roads, landscape areas etc.

GLOSSARY

Borel's σ–algebra Denote Borel's σ–algebra (which is generated by all open sets of a metric space) on $I\!R$ by $\mathcal{B}$ and on $I\!R_0^+$ by $\mathcal{B}_0^+$.

characteristic set function The characteristic function of a set is defined by

$$1_A(x) := \begin{cases} 1, & x \in A \\ 0, & x \notin A \end{cases}$$

Dirac measure Let $(E, \mathcal{E})$ denote a measurable space and $x \in E$ any element in it. The Dirac measure δ_x on x is defined by

$$\delta_x(A) := 1_A(x) = \begin{cases} 1, & x \in A \\ 0, & x \notin A \end{cases}$$

for all $A \in \mathcal{E}$, i.e. the whole mass is concentrated on the point x.

empty sum, empty product The empty set and the empty product shall always be interpreted as the neutral element of the additive and multiplicative element, respectively.

Gauss function The lower Gauss function $\lfloor . \rfloor$ is defined by

$$\lfloor x \rfloor := \max\{n \in Z\!\!Z : n \leq x\}$$

for all $x \in I\!R$. It yields the greatest integer that is less or equal to x.

identity kernel The identity kernel I on any measurable space $(E, \mathcal{E})$ is defined by $I(x, A) := 1_A(x)$, with 1_A denoting the characteristic set function.

intervals Intervals shall be written as follows:

$$\begin{aligned} [s, t] &:= \{u : s \leq u \leq t\} \\ [s, t[&:= \{u : s \leq u < t\} \\ ]s, t] &:= \{u : s < u \leq t\} \\ ]s, t[&:= \{u : s < u < t\} \end{aligned}$$

for all $s, t \in I\!R$.

kernel A (non–negative) kernel K from a measurable space $(E, \mathcal{E})$ to another measurable space $(F, \mathcal{F})$ is a function $K : E \times \mathcal{F} \to I\!R_0^+$ for which the following properties hold:

1 For any fixed set $A \in \mathcal{F}$, the function $x \to K(x, A)$ is measurable in $(E, \mathcal{E})$.

2 For any fixed $x \in E$, the set function $A \to K(x, A)$ is a measure on $(F, \mathcal{F})$.

A kernel K on a measurable space $(E, \mathcal{E})$ is a kernel from $(E, \mathcal{E})$ to $(E, \mathcal{E})$.

kernel iteration The nth iteration K^n of a kernel K on $(E, \mathcal{E})$ is defined recursively by $K^0 := I$, with I denoting the identity kernel, and $K^n := KK^{n-1}$ (see kernel multiplication) for all $n \in I\!N$.

kernel product Let K denote a kernel from $(E, \mathcal{E})$ to $(F, \mathcal{F})$ and L denote a kernel from $(F, \mathcal{F})$ to $(G, \mathcal{G})$. The multiplication KL is defined as the kernel from $(E, \mathcal{E})$ to $(G, \mathcal{G})$ with entries

$$(KL)(x, A) := \int_F K(x, dz)L(z, A)$$

for all $x \in E$ and $A \in \mathcal{G}$.

Kronecker function The Kronecker function δ is defined by

$$\delta_{xy} := \begin{cases} 1, & x = y \\ 0, & x \neq y \end{cases}$$

for any two elements x, y.

Kronecker product Let $A = (a_{ij})_{i,j}$ and $B = (b_{kl})_{k,l}$ denote any two matrices. The Kronecker product $C = A \otimes B$ is defined as the block matrix

$$A \otimes B = (a_{ij}B)_{i,j} = (a_{ij} \cdot b_{kl})_{(i,k),(j,l)}$$

numbers The set of positive integers is denoted by $I\!N$ and the set of non–negative integers by $I\!N_0$. The positive real numbers are denoted by $I\!R^+$ and the non–negative real numbers by $I\!R_0^+$.

projection Let I be an index set and $i \in I$. The projection mapping $pr_i : \prod_{k \in I} E_k \to E_i$ shall be defined as $pr_i(x) := x_i$ for $x = (x_k : k \in I)$. Thus the ith projection maps a vector x to its ith component x_i.

(sub–)stochastic kernel A non–negative kernel K from $(E, \mathcal{E})$ to $(F, \mathcal{F})$ is called sub–stochastic if $K(x, F) \leq 1$ for all $x \in E$ and stochastic if $K(x, F) = 1$ for all $x \in E$.

total variation Let μ denote a finite signed measure on $(E, \mathcal{E})$. Then there are (non–negative) finite measures μ^+ and μ^- such that the representation $\mu = \mu^+ - \mu^-$ holds. The norm of total variation on the space of finite signed measures is defined by $\|\mu\| := \mu^+(E) + \mu^-(E)$. Although the measures μ^+ and μ^- in general are not unique, the norm is independent of the specific choice and thus well–defined.

References

[1] A. Albert. Estimating the infinitesimal generator of a continuous time, finite state Markov process. *Ann. Math. Stat.*, 33:727–753, 1962.

[2] T. Anderson and L. A. Goodman. Statistical inference about Markov chains. *Ann. Math. Stat.*, 28:89–110, 1957.

[3] E. Arjas and T. Speed. Topics in Markov additive processes. *Math. Scandinav.*, 33:171–192, 1973.

[4] S. Asmussen. Phase-type distributions and related point processes: Fitting and recent advances. In: Chakravarthy and Alfa (eds.), Matrix-analytic methods in stochastic models, NY: Marcel Dekker. Lect. Notes Pure Appl. Math. 183, pp.137–149 , 1997.

[5] S. Asmussen. *Ruin probabilities*. Singapur: World Scientific, 2000.

[6] S. Asmussen and G. Koole. Marked point processes as limits of Markovian arrival streams. *J. Appl. Probab.*, 30(2):365–372, 1993.

[7] S. Asmussen, O. Nerman, and M. Olsson. Fitting phase-type distributions via the EM algorithm. *Scand. J. Stat.*, 23(4):419–441, 1996.

[8] S. Asmussen and H. Thorisson. A Markov chain approach to periodic queues. *J. Appl. Probab.*, 24:215–225, 1987.

[9] G. J. Babu and E. D. Feigelson. Spatial point processes in astronomy. *J. Stat. Plann. Inference*, 50(3):311–326, 1996.

[10] F. Baccelli and B. Błaszczyszyn. On a coverage process ranging from the Boolean model to the Poisson-Voronoi tessellation with applications to wireless communications. *Adv. in Appl. Probab.*, 33(2):293–323, 2001.

[11] F. Baccelli, K. Tchoumatchenko, and S. Zuyev. Markov paths on the Poisson-Delaunay graph with applications to routing in mobile networks. *Adv. in Appl. Probab.*, 32(1):1–18, 2000.

[12] F. Baccelli and S. Zuyev. Stochastic geometry models of mobile communication networks. In J. H. Dshalalow, editor, *Frontiers in queueing*, pages 227–243. CRC Press, Boca Raton, FL, 1997.

[13] N. Bambos and J. Walrand. On queues with periodic inputs. *J. Appl. Probab.*, 26(2):381–389, 1989.

[14] I. V. Basawa and B. Prakasa Rao. *Statistical inference for stochastic processes.* London etc.: Academic Press, 1980.

[15] H. Bauer. *Wahrscheinlichkeitstheorie.* Berlin etc.: Walter de Gruyter, 1991.

[16] D. Baum. Ein Faltungskalkül für Matrizenfolgen und verallgemeinerte Poisson–Gruppenprozesse. Research Report No.96–36, Dept. of Mathematics and Computer Science, University of Trier, Germany, 1996.

[17] D. Baum and V. Kalashnikov. Spatial generalization of BMAPs with finite state space. *J. Math. Sci., New York*, 105(6):2504–2514, 2001.

[18] A. Beck. On the strong law of large numbers. In F. Wright, editor, *Ergodic Theory*, pages 21–53. Academic Press, New York, 1963.

[19] R. Bellman. *Introduction to matrix analysis.* Philadelphia, PA: SIAM, 1997.

[20] R. F. Botta, C. M. Harris, and W. G. Marchal. Characterizations of generalized hyperexponential distribution functions. *Commun. Stat., Stochastic Models*, 3:115–148, 1987.

[21] R. J. Boucherie. Batch routing queueing networks with jump-over blocking. *Probab. Engrg. Inform. Sci.*, 10(2):287–297, 1996.

[22] R. J. Boucherie. On the quasi-stationary distribution for queueing networks with defective routing. *J. Austral. Math. Soc. Ser. B*, 38(4):454–463, 1997.

[23] R. J. Boucherie and N. M. van Dijk. Spatial birth-death processes with multiple changes and applications to batch service networks and clustering processes. *Adv. in Appl. Probab.*, 22(2):433–455, 1990.

[24] R. J. Boucherie and N. M. van Dijk. A generalization of Norton's theorem for queueing networks. *Queueing Systems Theory Appl.*, 13(1-3):251–289, 1993.

[25] R. J. Boucherie and N. M. van Dijk. On a queueing network model for cellular mobile telecommunications networks. *Oper. Res.*, 48(1):38–49, 2000.

[26] N. Bourbaki. *Elements de mathematique. Fasc. XXXV. Livre VI: Integration. Chap. IX: Integration sur les espaces topologiques separes.* Paris: Hermann & Cie, 1969.

[27] L. Breiman. *Probability.* Philadelphia, PA: SIAM, 1992.

[28] L. Breuer. An EM Algorithm for Batch Markovian Arrival Processes and its Comparison to a Simpler Estimation Procedure. *Annals of Operations Research*, 112. to appear.

[29] L. Breuer. Markovian Spatial Queues with Periodic Arrival and Service Rates. In Baum et al., editor, *Messung, Modellierung und Bewertung von Rechen- und Kommunikationssystemen. Kurzvorträge der 10.GI/ITG - Fachtagung MMB '99*, pages 97–102. Research Report No.99-17, Dept. of Mathematics and Computer Science, University of Trier, Germany, 1999.

[30] L. Breuer. Parameter Estimation for a Class of BMAPs. In G. Latouche and P. Taylor, editors, *Advances in Algorithmic Methods for Stochastic Models - Proceedings of the 3rd International Conference on Matrix Analytical Methods*, pages 87–97, New Jersey, USA, 2000. Notable Publications.

[31] L. Breuer. *Spatial Queues*. PhD thesis, University of Trier, Germany, 2000.

[32] L. Breuer. The Periodic BMAP/PH/c Queue. *Queueing Systems*, 38(1):67–76, 2001.

[33] L. Breuer. On Markov–Additive Jump Processes. *Queueing Systems*, 40(1):75–91, 2002.

[34] L. Breuer. Simple Solutions for a Class of Periodic Queues. In K. Amborski and H. Meuth, editors, *Modelling and Simulation 2002*, pages 505–512, Ghent, Belgium, 2002. SCS Europe.

[35] L. Breuer and D. Baum. The Inhomogeneous BMAP/G/∞ Queue. In B. Haverkort, editor, *Proceedings of the 11th GI/ITG Conference on Measuring, Modelling and Evaluation of Computer and Communication Systems*, pages 209–223, Berlin, 2001. VDE–Verlag.

[36] L. Breuer and M. Gilbert. Parameter Estimation for BMAPs. Research Report No.00–09, Department of Mathematics and Computer Science, University of Trier, Germany, 2000.

[37] D. Chauveau, C. Martin, A. van Rooij, and F. Ruymgaart. Discrete signed mixtures of exponentials. *Commun. Stat., Stochastic Models*, 12(2):245–263, 1996.

[38] Y. C. Chin and A. J. Baddeley. On connected component Markov point processes. *Adv. in Appl. Probab.*, 31(2):279–282, 1999.

[39] Y. C. Chin and A. J. Baddeley. Markov interacting component processes. *Adv. in Appl. Probab.*, 32(3):597–619, 2000.

[40] E. Cinlar. Markov additive processes and semi-regeneration. Proc. 5th Conf. Probab. Theory, Brasov 1974, 33-49 (1977).

[41] E. Cinlar. Markov additive processes. I. *Z. Wahrscheinlichkeitstheor. Verw. Geb.*, 24:85–93, 1972.

[42] E. Cinlar. Markov additive processes. II. *Z. Wahrscheinlichkeitstheor. Verw. Geb.*, 24:95–121, 1972.

[43] E. Cinlar. Markov renewal theory: A survey. *Management Sci., Theory*, 21:727–752, 1975.

[44] E. Cinlar. An introduction to spatial queues. In *Dshalalow, Jewgeni H. (ed.), Advances in queueing. Boca Raton, FL: CRC Press*, pages 103–118. 1995.

[45] E. Cinlar and H. Kaspi. Regenerative systems and Markov additive processes. Stochastic processes, Semin. Evanston/Ill. 1982, Prog. Probab. Stat. 5, 123-147 (1983).

[46] P. S. Cowpertwait. A generalized spatial-temporal model of rainfall based on a clustered point process. *Proc. R. Soc. Lond., Ser. A*, 450(1938):163–175, 1995.

[47] N. A. Cressie. *Statistics for spatial data.* New York etc.: John Wiley & Sons, 1991.

[48] D. Daley and D. Vere-Jones. *An introduction to the theory of point processes.* New York etc.: Springer-Verlag, 1988.

[49] M. Davis. *Markov models and optimization.* London: Chapman & Hall, 1993.

[50] A. Dempster, N. Laird, and D. Rubin. Maximum likelihood from incomplete data via the EM algorithm. Discussion. *J. R. Stat. Soc., Ser. B*, 39:1–38, 1977.

[51] P. J. Diggle. *Statistical analysis of spatial point patterns.* London - New York etc.: Academic Press, 1983.

[52] J. Doob. *Stochastic processes.* New York: Wiley, 1953.

[53] S. G. Eick, W. A. Massey, and W. Whitt. $M_t/G/\infty$ queues with sinusoidal arrival rates. *Manage. Sci.*, 39(2):241–252, 1993.

[54] S. G. Eick, W. A. Massey, and W. Whitt. The physics of the $M_t/G/\infty$ queue. *Oper. Res.*, 41(4):731–742, 1993.

[55] I. Ezhov and A. Skorokhod. Markov processes with homogeneous second component. I. *Theor. Probab. Appl.*, 14:1–13, 1969.

[56] I. Ezhov and A. Skorokhod. Markov processes with homogeneous second component. II. *Theor. Probab. Appl.*, 14:652–667, 1969.

[57] G. Falin. Periodic queues in heavy traffic. *Adv. Appl. Probab.*, 21(2):485–487, 1989.

[58] G. Fayolle, V. Malyshev, and M. Menshikov. *Topics in the constructive theory of countable Markov chains.* Cambridge: Univ. Press, 1995.

[59] W. Feller. *An introduction to probability theory and its applications. Vol II. 2nd ed.* New York etc.: John Wiley and Sons, 1971.

[60] S. G. Foss and S. A. Zuyev. On a Voronoi aggregative process related to a bivariate Poisson process. *Adv. in Appl. Probab.*, 28(4):965–981, 1996.

[61] I. Gihman and A. Skorohod. *The theory of stochastic processes II.* Berlin etc.: Springer-Verlag, 1975.

[62] I. Gikhman and A. Skorokhod. *Introduction to the theory of random processes.* Saunders, 1969.

[63] M. Gilbert. Zur Parameterschätzung für eine Klasse von Markov–additiven Prozessen. Diploma Thesis, University of Trier, 2000.

[64] J. Grandell. *Doubly stochastic Poisson processes.* Berlin etc.: Springer-Verlag, 1976.

[65] J. Harrison and A. J. Lemoine. Limit theorems for periodic queues. *J. Appl. Probab.*, 14:566–576, 1977.

[66] A. Hasofer. On the single-server queue with non-homogeneous Poisson input and general service time. *J. Appl. Probab.*, 1:369–384, 1964.

[67] H. Herrlich. *Topologie I. Topologische Räume.* Berlin: Heldermann Verlag, 1986.

[68] D. P. Heyman and W. Whitt. The asymptotic behavior of queues with time-varying arrival rates. *J. Appl. Probab.*, 21:143–156, 1984.

[69] J. Hofmann. *The BMAP/G/1 queue with level dependent arrivals: An extended queueing model for stations with nonrenewal and state dependent input traffic.* PhD thesis, University of Trier, Germany, 1998.

[70] J. Hofmann. The BMAP/G/1 Queue with Level–Dependent Arrivals - An Overview. *Telecommunication Systems*, 16(3-4):347–359, 2001.

[71] M. Jamshidian and R. I. Jennrich. Acceleration of the EM algorithm by using quasi-Newton methods. *J. R. Stat. Soc., Ser. B*, 59(3):569–587, 1997.

[72] N. P. Jewell. Mixtures of exponential distributions. *Ann. Stat.*, 10:479–484, 1982.

[73] V. V. Kalashnikov. *Mathematical methods in queuing theory.* Kluwer, 1994.

[74] E. Kamke. *Differentialgleichungen. Lösungsmethoden und Lösungen I.* Stuttgart: B.G. Teubner, 1977.

[75] S. Karlin and H. M. Taylor. *A second course in stochastic processes.* New York etc.: Academic Press, 1981.

[76] J. Kingman. *Poisson processes.* Oxford: Clarendon Press, 1993.

[77] G. Latouche and V. Ramaswami. *Introduction to matrix analytic methods in stochastic modeling.* Philadelphia, PA: SIAM, 1999.

[78] A. J. Lemoine. On queues with periodic Poisson input. *J. Appl. Probab.*, 18:889–900, 1981.

[79] A. J. Lemoine. Waiting time and workload in queues with periodic Poisson input. *J. Appl. Probab.*, 26(2):390–397, 1989.

[80] D. M. Lucantoni. New results on the single server queue with a batch Markovian arrival process. *Commun. Stat., Stochastic Models*, 7(1):1–46, 1991.

[81] D. M. Lucantoni. The BMAP/G/1 Queue: A Tutorial. In L. Donatiello and R. Nelson, editor, *Models and Techniques for Performance Evaluation of Computer and Communication Systems*, pages 330–358. Springer, 1993.

[82] D. M. Lucantoni, K. S. Meier-Hellstern, and M. F. Neuts. A single-server queue with server vacations and a class of non-renewal arrival processes. *Adv. Appl. Probab.*, 22(3):676–705, 1990.

[83] W. Massey. *Non–Stationary Queues.* PhD thesis, Stanford University, 1981.

[84] W. A. Massey and W. Whitt. Networks of infinite-server queues with nonstationary Poisson input. *Queueing Syst.*, 13(1-3):183–250, 1993.

[85] G. J. McLachlan and T. Krishnan. *The EM algorithm and extensions.* John Wiley & Sons, New York, NY, 1997.

[86] J. Mecke and D. Stoyan. The specific connectivity number of random networks. *Adv. In Appl. Probab.*, 33(3):576–583, 2001.

[87] B. Melamed and D. D. Yao. The ASTA property. In *Dshalalow, Jewgeni H. (ed.), Advances in queueing. Boca Raton, FL: CRC Press* , pages 195–224. 1995.

[88] X.-L. Meng and D. van Dyk. The EM algorithm - an old folk-song sung to a fast new tune. *J. R. Stat. Soc., Ser. B*, 59(3):511–567, 1997.

[89] S. Meyn and R. Tweedie. *Markov chains and stochastic stability.* Berlin: Springer-Verlag, 1993.

[90] E. Mourier. Éléments aléatoires dans un espace de Banach. *Ann. Inst. Henri Poincaré*, 13:161–244, 1953.

[91] J. Moyal. The general theory of stochastic population processes. *Acta Math.*, 108:1–31, 1962.

[92] W. Nef. *Lehrbuch der linearen Algebra.* Basel - Stuttgart: Birkhäuser Verlag, 1977.

[93] M. F. Neuts. Probability distributions of phase type. In *Liber Amicorum Prof. Emeritus H. Florin*, pages 173–206. Department of Mathematics, University of Louvain, Belgium, 1975.

[94] M. F. Neuts. Renewal processes of phase type. *Naval Res. Logist. Quart.*, 25:445–454, 1978.

[95] M. F. Neuts. A versatile Markovian point process. *J. Appl. Probab.*, 16:764–774, 1979.

[96] M. F. Neuts. *Matrix-geometric solutions in stochastic models.* Baltimore, London: The Johns Hopkins University Press, 1981.

[97] M. F. Neuts. The caudal characteristic curve of queues. *Adv. Appl. Probab.*, 18:221–254, 1986.

[98] M. F. Neuts. *Structured stochastic matrices of M/G/1 type and their applications.* New York etc.: Marcel Dekker, 1989.

[99] J. Neveu. Une generalisation des processus a accroissements positifs independants. *Abh. Math. Semin. Univ. Hamb.*, 25:36–61, 1961.

[100] J. Neveu. *Mathematische Grundlagen der Wahrscheinlichkeitstheorie.* München: R. Oldenbourg Verlag, 1969.

[101] P. Ney and E. Nummelin. Markov additive processes. I: Eigenvalue properties and limit theorems. *Ann. Probab.*, 15:561–592, 1987.

[102] P. Ney and E. Nummelin. Markov additive processes. II: Large deviations. *Ann. Probab.*, 15:593–609, 1987.

[103] E. Nummelin. *General irreducible Markov chains and nonnegative operators.* Cambridge University Press, Cambridge, 1984.

[104] S. Orey. *Lecture notes on limit theorems for Markov chain transition probabilities.* London etc.: Van Nostrand, 1971.

[105] A. Pacheco and N. Prabhu. Markov-additive processes of arrivals. In J. H. Dshalalow, editor, *Advances in queueing*, pages 167–194. CRC Press, Boca Raton, FL, 1995.

[106] M. J. Phelan. Point processes and inference for rainfall fields. In *Applied probability, Proc. Symp., Sheffield/UK 1990, IMS Lect. Notes, Monogr. Ser. 18, 127-148* . 1991.

[107] N. Prabhu. *Stochastic storage processes.* New York, NY: Springer, 1997.

[108] V. Ramaswami. A stable recursion for the steady state vector in Markov chains of M/G/1 type. *Commun. Stat., Stochastic Models*, 4(1):183–188, 1988.

[109] M.-A. Remiche. Asymptotic independence of counts in isotropic planar point processes of phase-type. *Adv. Appl. Probab.*, 32(2):363–375, 2000.

[110] M.-A. Remiche. On the exact distribution of the isotropic planar point processes of phase type. *J. Comput. Appl. Math.*, 116(1):77–91, 2000.

[111] M.-A. Remiche and G. Latouche. Asymptotic Poisson distribution in isotropic PH planar point processes. *Commun. Stat., Stochastic Models*, 16(2):259–272, 2000.

[112] T. Rolski. Approximation of periodic queues. *Adv. Appl. Probab.*, 19:691–707, 1987.

[113] T. Rolski. Queues with nonstationary inputs. *Queueing Syst.*, 5(1-3):113–129, 1989.

[114] T. Ryden. *Parameter estimation for Markov modulated Poisson processes and overload control of SPC switches. Statistics and telecommunications.* Lund Inst. of Technology, Dept. of Math. Statistics, 1993.

[115] T. Ryden. Parameter estimation for Markov modulated Poisson processes. *Commun. Stat., Stochastic Models*, 10(4):795–829, 1994.

[116] T. Ryden. An EM algorithm for estimation in Markov-modulated Poisson processes. *Comput. Stat. Data Anal.*, 21(4):431–447, 1996.

[117] T. Ryden. Estimating the order of continuous phase-type distributions and Markov-modulated Poisson processes. *Commun. Stat., Stochastic Models*, 13(3):417–433, 1997.

[118] R. Schassberger. *Warteschlangen.* Wien-New York: Springer-Verlag, 1973.

[119] R. Serfozo. *Introduction to stochastic networks.* Springer-Verlag, New York, 1999.

[120] H. Thompson. Spatial point processes, with applications to ecology. *Biometrika*, 42:102–115, 1955.

[121] D. Titterington. Recursive parameter estimation using incomplete data. *J. R. Stat. Soc., Ser. B*, 46:257–267, 1984.

[122] D. Titterington, A. Smith, and U. Makov. *Statistical analysis of finite mixture distributions.* Chichester etc.: John Wiley & Sons, 1985.

[123] G. J. Upton and B. Fingleton. *Spatial data analysis by example. Volume 1: Point pattern and quantitative data.* Chichester etc.: John Wiley & Sons, 1985.

[124] S. Verwijmeren, M. Mandjes, and R. J. Boucherie. Asymptotic evaluation of blocking probabilities in a hierarchical cellular mobile network. *Probab. Eng. Inf. Sci.*, 14(1):81–99, 2000.

[125] H. Willie. Individual blocking probabilities in the loss system $G_1 + \cdots + G_N/M/1/0$. *Queueing Syst.*, 6(1).109–112, 1990.

[126] H. Willie. Steady state of loss systems with a superposition of inputs. *Queueing Syst.*, 9(4):441–460, 1991.

[127] H. Willie. A note on single server loss systems with a superposition of inputs. *J. Appl. Probab.*, 34(1):213–222, 1997.

[128] H. Willie. Periodic steady state of loss systems with periodic inputs. *Adv. Appl. Probab.*, 30(1):152–166, 1998.

[129] R. W. Wolff. *Stochastic modeling and the theory of queues.* Englewood Cliffs, NJ: Prentice-Hall International, 1989.

Index

GPSR Compliance
The European Union's (EU) General Product Safety Regulation (GPSR) is a set of rules that requires consumer products to be safe and our obligations to ensure this.

If you have any concerns about our products, you can contact us on

ProductSafety@springernature.com

In case Publisher is established outside the EU, the EU authorized representative is:

Springer Nature Customer Service Center GmbH
Europaplatz 3
69115 Heidelberg, Germany

www.ingramcontent.com/pod-product-compliance
Ingram Content Group UK Ltd.
Pitfield, Milton Keynes, MK11 3LW, UK
UKHW021826190726
13853UKWH00003B/1208
* 9 7 8 9 4 0 1 0 0 2 4 0 0 *